Flight
Paths

Flight Paths

How a Passionate and Quirky Group of Pioneering Scientists Solved the Mystery of Bird Migration

Rebecca Heisman

HARPER

An Imprint of HarperCollins*Publishers*

HarperCollins books may be purchased for educational, business, or sales promotional use. For information, please email the Special Markets Department at SPsales@harpercollins.com.

FIRST EDITION

Designed by Kyle O'Brien

Library of Congress Cataloging-in-Publication Data has been applied for.

ISBN 978-0-06-316114-6

23 24 25 26 27 LBC 5 4 3 2 1

For #BirdTwitter, without whom
this book wouldn't exist;

and in memory of Bill Cochran, 1932–2022.

Contents

Flight
Paths

Introduction
Where Do the Birds Go?

I used to think I knew a lot about bird migration.

After all, I'd studied zoology in college, collecting data on the behavior of robins and sparrows for class projects and volunteering to help survey bird populations and monitor nest boxes. After graduation I'd found work as a seasonal field assistant on ornithology research projects on the prairies of Saskatchewan and in the Australian outback. I'd even kept a "life list" for a time, documenting more than six hundred bird species that I'd observed on my travels. Eventually I ended up working for the American Ornithological Society (AOS), the world's largest professional organization for scientists who study birds.

At AOS, a large chunk of my job was to publicize the research being published in its two venerable ornithological journals, which until recently were known as *The Auk* and *The Condor* (in 2021, these historic names were changed to *Ornithology* and *Ornithological Applications*). Instead of working in the field to help collect data, I spent my days at a desk reading cutting-edge migration research produced by others. I waded through scientific papers and exchanged emails with the researchers behind them as I translated their work into digestible blog posts, tweets, and press releases.

And for someone who thought I knew plenty about bird migration, I found myself being surprised an awful lot. Not just by what these scientists were learning—although that was fascinating, too—but by *how* they were learning it, the details in the sometimes-overlooked "methods" section of a scientific paper, where researchers spell out exactly how they produced their data. Despite my own background in ornithology, it was news to me that birds' migration patterns could be studied with weather radar. Or by analyzing the hydrogen isotopes in their feathers. Or with tiny devices that used the movement of the sun to calculate location. How, I wondered, did we even figure out how to *do* any of this? Almost every branch of science, it seemed, had been co-opted in service of figuring out the answer to one question: just where it is that birds go when they disappear south over the horizon in autumn.

Humans' curiosity about this goes back a very long time. Native American cultures seem to have figured out early that birds were flying away to distant locations when they vanished in the fall; Athabascan people in what is now Alaska, for example, have an old story about how Raven fell in love with a goose but had to part with her when fall arrived and she flew away over the ocean. European thinkers, however, took awhile to catch up.

Although some ancient Greek writers speculated that birds left for warmer locations, Aristotle threw things into confusion when he wrote his *Historia Animalium* in the fourth century BC. In it, he hypothesized that swallows hibernated in crevices in trees and that some winter and summer residents were in fact the same birds in different plumages—for example, that the common redstarts he saw in summer transmogrified into European robins when the seasons changed. Inspired by Aristotle, the Swedish priest Olaus Magnus suggested in the sixteenth century that perhaps swallows hibernated in the mud at the bottom of lakes and rivers, a misconception that persisted into the nineteenth century.

Perhaps the most outlandish idea, however, came from the English minister and educator Charles Morton. In the late seventeenth century, Morton, better remembered for writing a physics textbook that long remained in use at both Harvard and Yale, wrote a treatise in which he laid out his own fantastical theory of bird migration: they were simply flying to the moon. He estimated that if they could fly 125 miles per hour, it would take a flock of birds about two months to make the journey (although his approximation of Earth's distance from the moon was short by about 25 percent). As ridiculous as this sounds today, Morton was writing at a time when it was popularly believed that other planets must be inhabited and no one realized that there was a crucial lack of oxygen in the space between them. Some of the things he intuited about the natural history of migration turned out to be more or less right. He speculated that birds may be spurred to move to new areas by changing weather and a lack of food, and he even noted that body fat might help sustain them on their journey.

Early naturalists could guess all they wanted, but the first truly concrete evidence of where birds disappeared to every year arrived in the form of an unfortunate stork shot outside a German village in 1822. When the hunter went to pick up his prize, he must have been astonished to see that it had a massive spear impaled clear through its neck, which it had apparently been carrying around with it for some time. A German newspaper eventually analyzed the wood in the spear and its iron tip and concluded that it must have originated somewhere in Africa. Dubbed the *Pfeilstorch* (German for "arrow stork"), the bird was taxidermized—spear and all—and is still on display in a natural history museum in Rostock, Germany.

Since it had last departed Germany, the *Pfeilstorch* had not hibernated, transformed into a different species, or gone to the moon. Instead, it had been to Africa. Birds, it seemed, were traveling between continents.

Before we go any further, I should probably talk about what exactly bird migration is and how it came to be. Migration is simply the seasonal movement of animals between regions. Birds can be "permanent residents" that opt out of migration entirely to spend their whole lives in one place, short- or medium-distance migrants that move anywhere from a few miles up or down a mountainside to a few hundred miles, or—like most of the birds in this book—long-distance migrants, whose journeys span entire continents. Birds make these treks to take advantage of shifting resources at different locations throughout the year, chasing booms in the availability of insects and other key foods and the right conditions to nest and raise babies. The urge to migrate when spring and fall arrive can have a range of complicated triggers including changes in weather and day length as well as genetic programming. However it happens, though, there's a lovely German word that ornithologists use to describe this feeling that comes over birds: *Zugunruhe*, which literally means "movement restlessness."

Scientists have come up with two competing theories to explain how long-distance migration might have originated. The "northern home" hypothesis supposes that migrants are descended from birds that evolved at northern latitudes and eventually started to push farther and farther southward in the winter in search of milder climates. The "southern home" hypothesis is, just as you'd expect, the opposite—the idea that migrants started out as tropical birds looking north for better breeding grounds.

Analyzing the evolutionary family tree of long-distance migrants in the Americas suggests that the northern-home scenario was probably more common, and that some birds that live in the tropics year-round today are in fact the descendants of northern-home migrants that eventually began sticking around on their wintering territories permanently. Either way, striking out in search of better habitats made the ancestors of today's migrants more successful, they passed their wanderlust on to their offspring, and the evolution of long-distance migration was the result.

To make their epic voyages, birds rely on a range of navigation techniques. Genetic hard wiring seems to play a role, but migrating birds can also adjust on the fly (so to speak) by taking cues from the appearance of landmarks below and the orientation of the sun and stars above. Birds can even sense Earth's magnetic fields, through an inscrutable mechanism that recent research suggests may have something to do with quantum physics (!).

In the spring of 2020, vast new numbers of people discovered the magic of migration for themselves. Bird-watching, it turned out, was the ideal hobby for pandemic lockdowns. Visits to websites listing local bird sanctuaries and downloads of bird ID apps soared during those months. Retailers couldn't keep up with the increased demand for bird feeders and birdbaths, and people flocked (forgive the pun) to the Facebook groups of local bird-watching clubs.

The appeal was clear. You can watch birds almost anywhere, including, literally, your own backyard. The equipment needed to get started is minimal, and especially early on, when almost every species you learn to identify is new to you, it can provide a sense of novelty that's otherwise sorely lacking when you're stuck at home. And birds, after all, had never heard of coronavirus. Entering their world, if only for a little while, gave us a chance to forget what was going on in our own, and that first COVID wave in the United States happened to coincide with one of the most tantalizing phenomena in the natural world: spring migration. We might have been stuck at home, but birds of all shapes, sizes, and colors were traveling thousands of miles, returning from wintering grounds in Central and South America to their summer homes where they would find mates, build nests, and raise babies.

I was one of the many people paying closer-than-usual attention to the waves of migrants arriving that spring, enjoying seeing the cold and silent woods of my local birding patch filling anew with color and song. But like many, I was experiencing profound upheaval in both my personal and my professional life during those months. In need of a change, I decided it was time to turn some of what I'd learned at the American Ornithological Society into the proposal that became this book.

Many wonderful books have already been written on what migrating birds do and how they do it, but dig a little deeper, and you'll find there's another story here that's equally fascinating. We live in an era when you can go online and track the latest movements of an albatross via satellite nearly in real time, or download data on exactly how likely a given migratory species is to be present at any point on the globe, during any week of the year. How did we get from the *Pfeilstorch* to here?

The answer involves a sprawling group of ornithologists, engineers, and other scientists who've harnessed nearly every major technological development of the last hundred years in service of their quest to document bird migration in ever-greater detail, along with the legions of everyday bird-watchers documenting their observations of the world around them. In this book, I'll take you through techniques ranging from the origins of scientific bird banding to the latest breakthroughs in high-volume genetic sequencing. Along the way, I'll introduce you to some of the colorful characters who've made the science work—and, of course, the amazing birds whose journeys they've helped reveal. Whether you're a casual bird lover, a dedicated amateur ornithologist, or even a history of science buff looking for a fresh perspective, there is something in these pages for you.

If you've followed environmental news at all in recent years, you know that migratory birds are in trouble. But to save them, we need to know them. The people in these pages are the ones making that happen.

One

A Bird in the Hand

The summer I turned nineteen, I spent several weeks volunteering at Long Point Bird Observatory (LPBO), on the Canadian shore of Lake Erie. I'd just finished my sophomore year of college, where I was majoring in zoology, but this was my first hands-on experience with wildlife outside school. Officially, I was there to help with a decades-long research project on breeding tree swallows, which required monitoring hundreds of nest boxes and recording the birds' progress as they built nests, laid eggs, and raised babies. But in the mornings, I also got to participate in the work LPBO was best known for: bird banding.

Seven days a week, weather permitting, a dozen or so dedicated volunteers would rise before dawn to open the mist nets, lengths of fine nylon mesh strung like volleyball nets between metal poles. They were hung along paths through a small woodlot behind the building that housed researchers and volunteers, where birds would blunder into them as they went about their own morning routines, searching the little patch of forest for seeds or insects to eat. Mist nets secure birds in place without injuring them, and every twenty minutes we would walk along the nets to check for captured birds and then go through the delicate process of extracting them, working the loops of mesh free from their feet and wings.

I was there in time for the tail end of spring migration season, and every walk to check the nets felt like Christmas morning as I peered ahead to see what species we'd snagged. Would we spot the flame-orange throat of a Blackburnian warbler glowing in the net like an ember? The dapper black-and-white stripes on the face of a red-breasted nuthatch? Would a local cardinal have blundered in, ready to latch painfully onto the web of skin between my index finger and thumb with its massive orange bill? Extracting the birds was tricky work, and I never became very proficient at it. Woodpeckers in particular were notorious for sticking out their long, sticky tongues, which help them rake tiny insects out of bark crevices and wrap around the back of the birds' skulls when not in use, and getting them hopelessly entangled. But I loved patrolling the nets in the cool morning air, taking in the damp woodsy smell and the golden morning light filtering through the trees and the sound of birdsong.

Once they were out of the net, we would gently place the birds in cloth drawstring bags and carry them back to the banding lab. There, an experienced bander identified the species and used a special pair of pliers to close a numbered aluminum band around the bird's leg, reading out the number on the band to whoever was taking notes. A few moments were all it took to record a range of data about each bird, including its age and sex, some size measurements, and the amount of fat stored under the translucent skin of its belly, crucial fuel for the remainder of migration.

Weighing was always the last step. We weighed the birds by putting them headfirst into a small tube and then placing bird and tube together on an electronic scale, the immobilized bird's feet sticking comically out of the top. Finally, the bird was released directly from the tube via an open window in the lab. Watching a wren or sparrow or warbler wriggle out and shoot toward the trees, free again after its short interlude with us, was electrifying. Some of the birds we caught would be breeding nearby, but others still had a long journey ahead of them to reach nesting grounds in Canada's boreal forests.

A few of the birds we banded—not many, but a few—would be captured again, here or somewhere else, and the numbered bands would identify them, letting scientists track where they'd been and what condition they'd been in. I didn't fully appreciate it at the time, but the mornings I spent at the banding station that summer made me part of a scientific tradition stretching back more than a hundred years.

Why Band Birds?

Since its banding program began in 1961, the Carnegie Museum of Natural History's Powdermill Nature Reserve in southwestern Pennsylvania has banded more than 700,000 birds. Here are the recorded fates of a few:

Winter 1969–1970 or 1970–1971: A Swainson's thrush banded on September 24, 1966, is killed with a blowgun by an indigenous Peruvian roughly twenty-five miles south of the border with Ecuador.

October 8, 1971: A Traill's flycatcher banded on September 12 of the same year is found alive inside a bank in Belize City, British Honduras (now Belize).

May 22, 2001: A scarlet tanager identified as a "second year" bird when it was banded on June 21, 1990, is found dead near Houston, Texas. Just shy of twelve years old when it died, this bird holds the current longevity record for its species.

October 10, 2014: A ruby-throated hummingbird banded on September 18 of the same year is recaptured at a banding station in Lake Jackson, Texas. The distance between Powdermill and the Lake Jackson banding station is 1,425 miles, meaning the bird traveled an average of 65 miles every day for three weeks.

Powdermill is the longest-running year-round banding operation in the United States, and I knew that if I wanted to find out more about bird banding (aside from my own personal experience as a nineteen-year-old), they would have the answers. Annie Lindsay, the current manager of the bird banding program, started volunteering at Powdermill in 1999 when she was in high school, fell in love with bird banding, and never left. On top of her duties at Powdermill, she's also a PhD student, using banding data to look at how climate change is affecting the length of birds' wings.

Needless to say, Lindsay is busy, but she was kind enough to let me call her up and ask her all my bird banding questions on her one day off a week during fall migration.

The first thing I wanted to know was what exactly being the manager of such a large banding program takes. Unsurprisingly, it's no small job. Lindsay is in charge of making sure things are running smoothly at the banding station, coordinating with all of the volunteers and seasonal staff to confirm there are enough people there each day to be able to handle the volume of birds that they're expecting to get. "I'm also the person who trains new volunteers and field techs," Lindsay said, "and I process the birds," meaning she bands and measures them after they're removed from the nets.

Banding birds makes it possible for scientists to tell individuals apart; without a numbered band to go by, one robin or blue jay looks much like another. (Scientists who want to study the behavior of individual birds over a small area often add unique combinations of colored bands alongside the aluminum band so that, say, yellow-yellow-green can easily be distinguished from blue-yellow-yellow through binoculars.) In the United States, those bands are issued, and the resulting data archived, by the U.S. Geological Survey (USGS) Bird Banding Laboratory (BBL). As the Bird Banding Laboratory's website explains, "When banded birds are captured, released alive, and reported from somewhere else, we can reconstruct the movements of the individual bird."

Not just anyone can legally capture and band a bird. The Bird

Banding Laboratory only issues permits to people who have a specific research project in mind and have had training on how to do it safely, without injuring the bird. Aspiring permit holders submit a résumé of their past (supervised) bird banding experience and training along with their research plan and other information.

Those strict requirements help ensure that banding is overall very safe for birds. Although there have been a few reports over the years of metal bands causing leg irritation, the vast majority of problems that do arise come from human error when extracting birds from nets and traps, and even that is very rare. "Bird banders operate under a set of ethics, and our very first priority is the safety and well-being of birds," Lindsay said. Data from banding stations in the United States and Canada shows that for every thousand birds that are captured in mist nets, fewer than six suffer any sort of injury.

Mist nets, the primary tool banders use to capture birds, are usually about twelve meters long and two meters high, "made of really fine mesh that almost disappears if there's some nice vegetation behind it," Lindsay explained. Interspersed with the fine, loose mesh are heavier, tauter horizontal lines called shelves, which create pockets. "As the birds move through the habitat, they hit the net and drop into the pocket, and they're held there gently until someone can come around and extract the bird from the net."

Bands come in a range of sizes to perfectly fit every bird, snug enough to not slip over the bird's "ankle" and fall off but loose enough to spin freely and not constrict the bird's leg. There are tiny hummingbird bands that have an interior diameter about the width of a grain of rice and are stored on safety pins, and bands roomy enough to wear on your thumb for big birds like pelicans, swans, and eagles. A few birds, such as cardinals, receive special stainless-steel bands instead of aluminum; a cardinal's massive orange bill is powerful enough to tear off an aluminum band. As I discovered that summer at Long Point, getting bitten by a cardinal *hurts*.

The percentage of banded birds that are ever heard from again— recaptured at another banding station, spotted by a bird-watcher able to make out the numbers on the band, found dead, or shot by hunters—is tiny. According to Bird Banding Laboratory staff, it ranges from around one in ten for ducks and geese to one in four hundred for songbirds. But the BBL distributes about a million bands every year, and those small percentages add up.

And banding is useful for studying much more than just migration. When I asked Lindsay what ornithologists can learn from bird banding, she almost didn't know where to start. "There's so, so much," she said. Banding birds can help scientists figure out how long birds live, how large their populations are, how they behave when defending a territory and raising young, and much more. Banding also has a special role in the management of waterfowl populations, because it helps wildlife authorities set harvesting limits for ducks and geese; there are even special "reward bands" that hunters receive a small cash reward for reporting, to help calculate the chances that someone who shoots a banded bird will report it. But banding got its start with ornithologists looking to the sky and wondering where the birds were going, and it's helped us answer that question, too.

Banding records, for example, helped scientists track how the migratory behavior of house finches changed after they were introduced into eastern North America in the 1940s. This familiar backyard bird, which resembles a sparrow that dipped its face in red paint, is native to the western United States, where only a tiny fraction of individuals move around between seasons. When a population became established in the East, however, their behavior changed quickly. Two birds banded north of New York City in 1959 turned up in the Philadelphia area and then, the following spring, back in the spot where they were originally captured, showing they'd made at least a short round-trip.

And they didn't stop there. In 1998, the researcher Kenneth Able analyzed all the available bird banding data up to that point on house finches captured in the eastern United States and Canada and later reencountered. He showed that within twenty years of the birds' introduction in the East, between a quarter and half of all house finches in the eastern population were migrating each year, despite being descended from birds that were almost all homebodies, and the number of birds that were migrating and the distance they were traveling was continuing to increase, with some flying all the way to Florida and the Gulf Coast in search of a mild winter climate. In fewer than sixty generations, Able suggested, the different pressures the birds were subjected to in their new habitat had led to the rapid evolution of migratory behavior.

Banding records also provided the first hints that something strange and wonderful was going on with the migratory route of a tiny bird called the blackpoll warbler. These birds (whose name comes from the black caps on the tops of their heads) nest as far northwest as Alaska, and to get there from their South American wintering grounds in the spring, they make a crossing to Florida and fan out from there. The obvious assumption would be that they do the reverse in the fall, but in 1970 the ornithologist Ian Nisbet pointed out that this didn't line up with what bird banding stations were actually observing every autumn.

If the birds headed south via the same route they took north, banding records from the East Coast should have shown their progression south as the fall migration advanced. Instead, when Nisbet analyzed the records, he was surprised to find that blackpoll warblers were "progressively *less* numerous" at banding stations farther and farther down the coast in the fall. In Florida, capturing one in the fall was actually quite rare. Nisbet suggested that the birds were bypassing the region entirely in the autumn and doing something that seemed impossible for a bird that weighs about as much as a ballpoint pen: flying directly from New England to northern South America, crossing fifteen hundred miles of open ocean.

It would take another four decades and a tracking device based on navigation principles dating to the Renaissance to prove Nisbet right. But long before any of this research could occur, someone had to have the idea to put tiny bracelets on birds and see what happened.

Banding's Origin Story

John James Audubon, namesake of the Audubon Society, is often credited as the father of bird banding. In his autobiography, he described tying silver threads to the legs of five eastern phoebe nestlings in Pennsylvania in the spring of 1804, when he was nineteen years old—the same age I was during my summer at Long Point.

"These they invariably removed," he wrote, "either with their bills, or with the assistance of their parents. I renewed them, however, until I found the little fellows habituated to them; and at last, when they were about to leave the nest, I fixed a light silver thread to the leg of each, loose enough not to hurt the part, but so fastened that no exertions of theirs could remove it."

The following year, as he told it, he spotted two phoebes bearing his silver threads on the same property, returned after their annual migration. Audubon had demonstrated for the first time that migratory birds return to the area where they were fledged. Or so he said.

It turns out that young Audubon was actually in France at the point in the spring of 1805 when he claimed he'd resighted his marked phoebes. And modern studies have found that only 1 to 2 percent of eastern phoebes return to the site where they were raised to breed, making his claim that two out of his five birds returned home extraordinary. Matthew Halley, the biologist and historian whose sleuthing uncovered these discrepancies, speculates that Audubon probably did attach the threads to the nestlings' legs and just made up the bit about seeing them again the next year, part of an apparent habit on Audubon's part to embellish the occasional fact in ways "favorable to his legacy." America's most famous ornithologist, apparently, liked to stretch the truth.

But even if Audubon was guilty of employing some alternative facts when he wrote his life story, the idea of marking individual birds to learn more about their movements eventually caught on. In the 1880s, a naturalist named Ernest Thompson Seton marked the breasts of snow buntings on his farm in Manitoba with printer's ink to see whether the same birds stuck around all winter. (Apparently not, because he never saw his inked birds again. Seton would go on to greater fame as one of the founders of the Boy Scouts of America.)

The scientific use of metal bands originated in Europe in 1899 with a Danish ornithologist named Hans Christian Cornelius Mortensen, who worked first with starlings before moving on to ducks, storks, and birds of prey. Early bands like Mortensen's often had the initials of the bander or location engraved on them for identification.

Banding (known as "ringing" in Great Britain) quickly took off both in Europe and elsewhere, and in the United States the story picks up with Leon J. Cole. Cole was born in 1877 and raised in Albany, New York. Although a "city boy" in his youth, "L.J." nevertheless developed an early interest in plants and animals. He spent his summers working on farms before going off to college, and eventually he received a bachelor's degree in biology from the University of Michigan and a PhD in zoology from Harvard.

While he was a student in Michigan, Cole was inspired by a U.S. Fish Commission project releasing live, tagged fish and asking fishermen to report them as a means of learning about their movements. At this point, Cole had never heard of either Audubon's (questionable) experiment with phoebes or Mortensen's efforts in Europe, but he wondered if something like the fish tagging program might be possible with birds. In 1901, at the age of twenty-four, he gave a talk to the Michigan Academy of Science titled "Suggestions for a Method of Studying the Migrations of Birds."

"It is possible such a plan might be used in following the movements of individual birds," he said, "if some way could be devised of numbering them which would not interfere with the bird in any way, and would still be conspicuous enough to attract the attention of any person who might chance to shoot or capture it."

For six years, it remained just an idea. But in 1907, Cole became a zoology instructor at Yale University, where he conducted research on the genetics of pigeons. It was there, as a member of the New Haven Bird Club, that he finally got the chance to put his plan into action.

By now, Cole was aware that he wasn't the first person to think up the concept of bird banding, having heard about Audubon's and Mortensen's work. He didn't get to be the first person to actually try out bird banding in the United States, either; two others had beaten him to it since his original proposal in 1901. In 1902, a Smithsonian scientist named Paul Bartsch had experimented with banding young black-crowned night herons in the Washington, D.C., area, one of which turned up dead in Cuba two years later, the first long-distance banding "return" in North America. And in 1904, P. A. Taverner of Detroit distributed about two hundred aluminum bands to correspondents via a notice he placed in *The Auk*, the journal of the American Ornithologists' Union (AOU). One of Taverner's bands was used in Iowa to tag a northern flicker (a type of woodpecker) that was later recovered in Louisiana.

But Cole had grander ambitions for American bird banding. He realized that for the data collected to really be useful, some sort of centralized record keeping would be essential, rather than a hodgepodge of small individual efforts. (In one of his early papers on the subject, he noted the recovery of two ducks wearing bands inscribed with the initials TJOD; who or what TJOD was, no one ever figured out, making their recovery meaningless for science.) Toward that end, in the winter of 1907–1908 Cole and his New Haven comrades developed a system using numbered aluminum bands stamped with the club's mailing address, along with standardized forms for recording information about each banded bird.

The number of birds tagged that first spring was "disappointingly small"; the plan was that club members would carry bands with them on their field trips and simply use them whenever the opportunity arose, primarily on baby birds still in the nest, but apparently "the chief interest of the field workers [was] along other lines." Cole, however, was undeterred, and in 1908 the New Haven Bird Club scaled up its efforts under his direction. Now using bands stamped "Notify The Auk, New York," it distributed around five thousand of them to ornithologists around the country.

About one thousand of those bands were used, and by the time Cole reported back to the American Ornithologists' Union in December 1909, 3 percent of the banded birds had been "heard from since." The most commonly banded bird, he reported, was the American robin, and their first "return" was of a robin banded in Rhode Island in 1908 and "taken" (presumably shot) the following spring in the same vicinity. None of the recovered birds showed any signs of harm from being banded.

Cole and his colleagues received some interesting letters from people who found their banded birds. Here's one he reproduced in his 1922 account of this work, sent with a report of a banded night heron, with the original spelling, grammar, and capitalization preserved:

Gentlem dear sirs Your bird was shot here to day by me Albert Bailey for which I was more than Sorry when I found it had a ring on. I took it for a Hawk as It flew several times over my yard as I thought after chickens and Gentlemen all I can say that I am sorry If I did wrong In so doing and also beg Pardon.

> Yours with Rees,
> Albert Bailey

Managing this expanded banding program while also teaching zoology at Yale proved a bit much for Cole to handle, though. Distributing bands far and wide "increased greatly the burden of correspondence, the difficulties of securing bands, and of obtaining funds to meet necessary expenses," he wrote in 1922. The correspondence and record keeping "fell entirely to myself and had to be conducted without clerical assistance," which, as he mildly put it, "grew to be a considerable burden." Cole and his bird club buddies were even making most of the bands themselves, cutting up sheets and tubes of aluminum and stamping the bands with the numbers and contact information before sending them out.

Keeping up with the bird banding program on his own wasn't going to be sustainable, and on December 8, 1909, Cole gave another presentation on his efforts so far to an American Ornithologists' Union meeting in New York City in which he pleaded for the establishment of a formalized, permanent organization to oversee bird banding in the United States. At dinner that night, the American Bird Banding Association was formed, with Cole as its president.

As Cole's career in the new field of genetics took off, however, his involvement in bird banding faded. In April 1910, Cole moved from Connecticut to Wisconsin to head up the University of Wisconsin's newly formed Department of Experimental Breeding, the forerunner of its genetics program. He muddled along as best he could with the American Bird Banding Association for one more year until, about to leave for a summer in Europe, he had to face the fact that someone else would need to take over. He arranged for the Linnaean Society of New York City, an organization of professional and amateur naturalists founded in 1878, to take over for the AOU in overseeing the association, and Cole bowed out. This arrangement lasted until 1920, when bird banding became the responsibility of the federal government's Bureau of Biological Survey, forerunner of the U.S. Fish and Wildlife Service (USFWS).

When the federal government took over the oversight of bird banding in 1920, the fledgling North American Bird Banding Program was headed up by the ornithologist Frederick C. Lincoln, who would hold the job for more than twenty years and become a legend in the field for his work with waterfowl banding records. Leon Cole, on the other hand, never really returned to his own initial interest in bird migration. Unfortunately—but perhaps unsurprisingly, for a scientist working in the early twentieth century and running a Department of Experimental Breeding—Cole later became involved in another field: eugenics.

This movement, which advocated for improving the human race through selective breeding, found broad support among American biologists, sociologists, politicians, and others in the first half of the twentieth century. Today, though, it's impossible to view it as anything but the basest racism dressed up in pseudoscientific clothing. American eugenicists successfully lobbied to severely curtail immigration from nonwhite countries and pass antimiscegenation laws to prevent interracial marriage. Hitler drew inspiration from the American eugenics movement when hatching his plans for what became the Holocaust. Science, as history has shown time and again, can all too easily be misapplied by humans with prejudices. Cole gave speeches advocating for the "permanent segregation" of "defectives and cripples" to prevent them from breeding and passing on their inferior qualities and acted as a scientific consultant on the eugenic benefits of birth control.

Today Cole is mostly remembered neither for his involvement in bird banding nor for his support of eugenics, but for his work on the genetics of pigeons, work he began at Yale and continued in Wisconsin. Late in life, he mentioned to his friend Robert McCabe that he'd always wanted to be named a fellow of the American Ornithologists' Union—a special honor conferred upon members for their "exceptional and sustained contributions to ornithology." McCabe put together the nomination paperwork and sent it off.

But they never heard anything back. "What happened, I never knew, but the AOU turned him down—forgotten even by the enlightened and those who should have remembered," wrote McCabe. "[His] disappointment was obvious, but never discussed between us." It was 1946, the year of the Nuremberg trials, and the world was becoming aware of the full horrors of the concentration camps. I'd like to think that Cole's more unsavory interests had caught up with him at last.

Sparrow Traps and Mist Nets

Cole and his contemporaries were mostly putting bands on baby birds that hadn't left the nest, simply because they were the easiest to catch. Safely trapping adult birds to band them was a challenge that took some time to unravel.

One of the first to tackle it was a rakish Canadian named Jack Miner, or "Wild Goose Jack." After establishing a waterfowl sanctuary on his property in Ontario in the opening years of the twentieth century, he became determined to figure out where the ducks and geese that passed through every year were headed. Eventually, he devised a trap for Canada geese that consisted of a canal connecting two ponds with a trapdoor at either end and caught his first goose in the spring of 1915.*

All that was known at that point about the migration of Canada geese, his son Manly (who carried on his work) later wrote, "was that they went north, and the settler in the most northern point in Canada always reported that they went still farther north." But that October, Miner received a letter from a representative of the Hudson Bay Company, informing him that his banded goose had been killed by "an Indian" in August in the vicinity of Hudson Bay, where it had been nesting. Miner would go on to band thousands of geese and ducks, vastly improving ornithologists' understanding of their routes to and from their Arctic breeding grounds.

Catching smaller birds, like songbirds, required a different approach. The first person in the United States to attempt it on a large scale was a lawyer from Ohio. "Never of very rugged constitution even from early youth," according to later accounts, Samuel Prentiss Baldwin had retired from practicing law due to poor health in 1902, when he was only thirty-four. He pursued other businesses, but he also took an interest in the birds on his farm outside Cleveland.

Around 1913, he became determined to eradicate house sparrows from the property. Introduced from Europe about sixty years before, this species had quickly spread across North America, competing with

---

* In addition to his contact information, Miner's bands were stamped with short Bible verses such as "Have faith in God." His descendants still band geese in Ontario using similar biblical bands, and "Miner bands" are sought-after trophies for waterfowl hunters.

native birds for food and nest sites. To get rid of them, Baldwin used a contraption called a "Government Sparrow Trap," promoted by the U.S. Biological Survey for getting rid of the pesky invasive sparrows. Consisting of a wire mesh cage into which birds were lured with a bait of grain, the trap apparently did its job. "The Sparrows were destroyed in large numbers, and the farm pretty well cleared of them, greatly to the comfort, evidently, of the native birds," wrote Baldwin, "for it was very noticeable that, as the Sparrows decreased in number, the native birds greatly increased."

The traps captured the birds alive; the "destroying" had to be carried out separately. And when Baldwin heard about the recently formed American Bird Banding Association, he began to view them in a new light. As the number of house sparrows on his farm decreased, more and more native birds had been showing up in his traps, and in 1914 he began banding them.

The photos included with his treatise on his trapping and banding efforts, published by the Linnaean Society of New York in 1919, show Baldwin operating traps and handling birds while wearing a bow tie and a dapper boater hat. By that time, he had banded nearly sixteen hundred birds and had received outside reports about the fates of three, including a robin that turned up in South Carolina. He'd also devised some additional trapping methods of his own, such as using trapdoors to capture birds in nest boxes.

Some birds returned to his baited traps again and again in search of grain. "One learns to know the characters of certain individuals, as I came to know a certain White-throated Sparrow, who always identified himself by fighting and biting my fingers," he wrote, "and another White-throat, who distinguished himself as a squealer."

Baldwin died in 1938 at the age of seventy. Variations on his trapping methods continued to be the standard for bird banders in the United States until the introduction in the 1950s of the piece of gear I helped use during my summer at Long Point: the mist net.

Fine nets made of black silk, strung between bamboo poles, had been used by the Japanese since the seventeenth century to catch small birds as they made their way from breeding grounds in Siberia and northeastern Asia to wintering grounds in southern Japan. These original mist nets weren't part of any scientific research, however; the birds caught in them were eaten. Similar nets were used for the same purpose in parts of southern and eastern Europe, and American ornithologists had stumbled across immigrants using them in California as early as the 1920s. But mist netting didn't really start to take off with scientists in the United States until 1947, when Oliver Austin, a Harvard-educated ornithologist who headed up wildlife management in Japan under the Allied occupation following World War II, described them in one of his reports from the island nation.

Mist nets soon caught on in the United States with bat researchers as well as ornithologists, although figuring out exactly how best to use them took some trial and error. In 1950, the Louisiana State University zoologist George Lowery (remember that name—Lowery and his students pop up repeatedly in the history of migration research) gave the mammalogist Walter Dalquest three Japanese silk mist nets to try out in his work inventorying bat species in the Mexican state of San Luis Potosí. In his account of the attempts that followed, Dalquest never gives the name of the Mexican field assistant whose suggestions about net placement led to their catching their first bats. In addition to all the awkward crashing around in the brush at night necessary to net the night-flying animals, however, Dalquest described an incident in which a "wandering cow" blundered into one of the nets and entangled its horns. "The net had slipped from its supporting poles and the cow, with my men after it, had crashed off through the brush," wrote Dalquest. After they finally recovered the net, "the departing cow was warned with a load of dust shot never to return."

Once scientists got the hang of them, the advantages of mist nets over baited wire traps became obvious. For one thing, the only birds attracted to traps baited with grain were birds that ate grain; mist nets could ensnare a much richer variety of species, including the many that preferred insects. For another, as long as the person carrying out the netting knew what they were doing, they seemed to be safer for the birds themselves. Birds in hard-sided traps were prone to injure themselves by flapping ceaselessly against their restraints, and because traps were typically checked less frequently, birds captured in them were often vulnerable to passing predators.

Mist nets aren't the only method used today to capture birds for banding; cannon nets and rocket nets fire large nets out to ensnare groups of shorebirds on beaches, for example, and descendants of Baldwin's baited traps and nest box trapdoors are still used in some situations. But modern mist nets, which are made of nylon or polyester and come in a range of mesh sizes for use with different bird species, have become inseparable from the practice of bird banding.

No matter how a bird is caught, however, the real magic begins when someone gets the chance to take an intimate look at it and see what they can learn.

The Bird Banding Laboratory

"Having a bird in your hand—it just changes everything." Tony Celis-Murillo's voice turns warm and soft as he talks about what it's like to band a bird. "Learning the details of identification and things like that—it's great with binoculars, but still it's far away. When you have it in your hand, and you can look at the details, it just engages you right away."

When we spoke in the fall of 2020, Celis-Murillo was the acting chief of the USGS Bird Banding Laboratory, filling in for the permanent chief, who retired in 2019. A native of Mexico, Celis-Murillo volunteered with fish and mammal biologists as an undergrad at the Universidad Autónoma del Estado de Morelos, but it was when he got his first taste of bird banding that he fell in love. Much of his research on birds' song and behavior has relied on marking birds with colored bands to make individuals easily identifiable as they go about their lives.

"It was so cool doing those observations, spending hours in the field and seeing, oh, red-green-yellow-aluminum is fighting with green-green-yellow-aluminum. You start to get to know everyone, and you talk about the color combinations like names," he says. "Oh, remember green-green-yellow? Yeah, he's always a mess!"

The Bird Banding Laboratory has been shuffled among various federal agencies over its hundred-year history, becoming part of the USGS after a reorganization in 1996. (It jointly oversees the North American Bird Banding Program with its Canadian counterpart, the Canadian Bird Banding Office.) While the actual process of putting a band on a bird has changed little in the last century, the technology used to collect and archive the resulting data has gone through a series of monumental shifts.

For most of the BBL's history, bands were inscribed with a mailing address, just as Cole's were when he began his first centralized distribution of bands. People who found a banded bird (usually members of the public stumbling across a dead bird, or hunters shooting a banded duck or goose) would write to the lab to report their observation. In the 1990s, the BBL switched to bands stamped with a toll-free phone number and started staffing a call center to receive reports. Either way, anyone who submitted a report would receive a certificate of appreciation with details about their bird, including where and when it was banded and its approximate age at the time.

"We have stories of people who have walked in the forest of Central America for a week to find the closest post office, because they wanted to know where this bird came from and what the band meant," says Celis-Murillo. The certificates of appreciation distributed by the BBL engaged the imaginations of thousands of people. Imagine learning that the tiny gray-and-yellow warbler that turned up dead in your driveway in Texas had been captured and tagged by scientists two years before in Ontario, Canada, and had likely been on its way to South America when felled by bad weather and exhaustion: no matter how sad you felt for the bird, your discovery made you part of a network of researchers studying one of the most amazing natural phenomena on the planet.

In 2006, the BBL launched the website reportband.gov, and bands issued today bear a URL instead of an address or phone number. About 100,000 "encounter reports" are submitted each year, and the number is growing. In addition to people finding and reporting dead banded birds, there are more people taking up bird-watching as a hobby than ever before, and many of them carry powerful spotting scopes and cameras into the field that let them make out band numbers without the need to recapture birds. "It seems like we're going to achieve this countrywide support of eyes and hands out there who are contributing to learning more about birds," says Celis-Murillo.

Because bird banding also serves as the basis for many other, higher-tech bird research techniques, BBL staff also have to keep abreast of the latest cutting-edge methods for tracking birds—even as they remain the caretakers of the oldest method of all.

"Just for example, there is a new way of tracking now from the International Space Station, Icarus," says Celis-Murillo. "It's great, we're going to have more precise data, we're going to learn a lot. But in our database, we want to know how many people are using Icarus on banded birds, on what species, all of that. So we need to be sure our database now has a new field, a new code, for Icarus transmitters." We'll return to Icarus transmitters in a later chapter.

Efforts to digitize the decades of data collected by the BBL began in the 1990s. So far, they've made it as far back as the 1960s, but they're not done yet; in addition to simply being scanned into a computer, old reports and data sheets must be carefully checked for errors to make sure the information they contain will be as useful to future scientists as possible. And as birds' migratory behavior begins to shift in response to climate change, this long-term record is becoming more important than ever.

A New Use for Old Data

By their very nature, long-term data sets—records that track a specific phenomenon in a consistent way across decades—often end up being used to answer questions that didn't even exist when data gathering began. Scientists develop new techniques for statistical analysis that let them tease out patterns in the data that previous generations of researchers couldn't see, and long-term trends begin to appear against the static of normal year-to-year ups and downs.

Leon Cole and Samuel Baldwin didn't have climate change on their minds when they started capturing and banding birds in the early years of the twentieth century. The phrase "global warming" wouldn't even be coined until decades later. But the system they helped put in place has provided us with an unexpectedly rich portrait of how our shifting climate is affecting birds' delicately timed journeys. At this timescale, a banded bird doesn't even need to be recaptured to provide useful data; just having a long-term record of when birds of different species pass through banding stations each year is enough.

In 2009, a group of researchers led by Josh Van Buskirk analyzed Powdermill banding data going back to 1961—almost 400,000 individual records representing seventy-eight species—to look for trends in the timing of spring and fall migration. Although the details varied between species, overall they found that spring migration begins earlier and takes longer now than it did decades ago. It's more complicated than you might think to point a finger directly at climate change, but their results did show that birds migrated earlier in warmer years. Banding data from Europe shows that there, too, birds have been arriving on their breeding grounds earlier and earlier in recent decades.

This left Buskirk and his colleagues wondering how exactly these changes in the timing of bird migration are happening. Is it simply a matter of what biologists call "phenotypic plasticity," the flexibility that lets an individual organism adjust its behavior in response to changes in its environment, or could something more be going on? Here, too, banding data can help provide answers. For a follow-up study in 2012 using the same data set from Powdermill, Buskirk and his colleagues were able to estimate birds' phenotypic plasticity by looking at how well the arrival timing of individual birds that were captured in more than one year matched up with year-to-year changes in temperature. It turned out that this individual flexibility couldn't explain more than a quarter of the overall decades-long trend toward earlier migration shown by the data. The rest, the scientists concluded, must be "micro-evolution" over the twenty generations or so that would have passed during the time span of the data set.

More recently, scientists have used banding data to examine how the timing of migration for a single species has been changing on a continental scale, instead of how migration for many species has changed at a single location. Between 2016 and 2020, the research-ers Sara Morris and Kristen Covino worked together on a series of studies that showed that blackpoll warblers' fall journey (that amazing flight over the open Atlantic) has been happening about one day later per decade since 1967, while their spring migration along a different route has been getting about half a day earlier each decade. A similar analysis for a related species, the black-throated blue warbler, showed that its spring migration has also been advancing by about a day per decade.

If this all sounds like good news—the data shows that birds can adapt to climate change!—unfortunately it's not that simple. There are limits to adaptation, and we don't know how the rate of migratory birds' evolution will compare with the rate of environmental change in the future. What's more, it appears that birds are adjusting their annual schedules at a different pace from the insects, plants, and other organisms they rely on. This is throwing them into a situation called phenological mismatch: if birds are arriving on their nesting grounds one week earlier than they used to, but the short-lived caterpillars they feed their babies are hatching three weeks earlier than they used to, that's a problem. Populations of many migratory bird species across North America are already seeing sharp declines.

"One of the really good things about bird banding is the oppor-tunities it provides for outreach. When people get to see a bird up close—most of them leave just amazed, and just from being there for half an hour, they've learned so much," Powdermill's Annie Lindsay told me. "Understanding is the key to conservation. If they start to care about this organism, then they start to care about conserving it. That's one of the things that I really feel strongly about."

I know what she means.

In February 2019, almost fourteen years after that summer in Long Point, my family and I attended a celebration of winter birds at McNary National Wildlife Refuge near our home in southeastern Washington. Bundled up against the chilly weather, my husband, our almost-two-year-old son, and I walked around the refuge head-quarters, a short distance from the broad, flat expanse of the Columbia River. There were live raptors being shown off by staff from a nearby wildlife rehabilitation center, a bird blind from which we could watch the snow geese on the nearby slough, and a bird banding demonstration.

Mist nets snaked through the brush, some of them paralleling the path to the blind. Here and there we could spot a bird hanging helplessly in a net, fluttering a bit as it waited for a trained volunteer to extract it for its turn to be banded. Most of them were white-crowned sparrows, little brown birds with bold black-and-white helmets, which winter here but in a couple more months would return to breeding grounds in Canada to the north or the Rocky Mountains to the east.

We joined a crowd of children and their parents around a folding table that had been turned into a temporary banding station, with data sheets, field guides, a digital scale, strings of bands, pliers, all the ac-coutrements of a bander. A man bent his head over the sparrow gently gripped in his hand, blowing softly on the downy feathers on its belly to check for fat deposits under the skin, examining the details of its plumage to estimate its age.

The whole process, from removing the bird from its cloth bag to releasing it, took less than a minute. Freed, the sparrow shot back into the bushes to collect itself, but the crowd lingered, wanting to see another bird, and another. Each one was another data point in our understanding of birds' lives and movements and how our shifting climate is affecting the world we share with them. But each one was an ambassador, too, its minute in the limelight helping to bridge the gap between species. Who knows—maybe one of the kids watching (maybe even mine!) would grow up to be an ornithologist.

Two
Looking and Listening

A chill wind was blowing when the historian and amateur orni-thologist Orin Libby climbed a hill west of Madison, Wisconsin, on the evening of September 14, 1896. But however loud the wind was, the calls of night-migrating birds passing overhead were louder: over the course of five hours, Libby counted thirty-eight hundred birdcalls, an average of twelve per minute.

"The air seemed at times fairly alive with invisible birds as the calls rang out, now sharply and near at hand, and now faintly and far away," wrote Libby when he published his observations. "Almost human many of them seemed, too, and it was not difficult to imagine that they expressed a whole range of emotions from anxiety and fear up to good-fellowship and joy.... It was a marvel and a mystery enacted under the cover of night, and of which only fugitive tidings reached the listeners below." You don't find writing like that in scientific journals anymore, and that's a pity.

Most birds migrate at night, rather than in daylight. Traveling by night has multiple advantages: fewer predators are out and about looking for a feathery snack; if it's clear, the moon and stars can help with navigation; and the air tends to be less turbulent, making for smoother flying.

But for ornithologists, nocturnal migration has an obvious downside: it's hard to see what the birds are actually up to. Banding helps, but trying to grasp the ebbs and flows of bird migration across entire continents based on a handful of banding records is like trying to reconstruct the scene depicted on a tapestry by following a few individual threads. Before the advent of high-tech tracking devices and other modern techniques for studying migration, however, ornithologists had already come up with some clever work-arounds using only their eyes and ears. We'll come back to the part about eyes in a bit (hint: the full moon is involved), but on that autumn night in 1896, Libby had stumbled upon a powerful tool for glimpsing the full scale of migration—nocturnal flight calls.

Songbirds, cuckoos, and woodpeckers are among the birds that vocalize regularly during their long night flights, often producing calls that are very different from the ones bird-watchers might be familiar with from watching the same species during the day. Ornithologists' best guess as to *why* is that these calls help keep groups together and stir the urge to migrate in other birds that hear them. Recent research shows that warbler species following similar migratory routes have similar nocturnal flight calls regardless of how closely related they are, suggesting that flight calls might help birds taking the same route find one another and stick together. For example, the blackpoll warbler we met in the last chapter isn't the only bird that launches itself out over the open Atlantic every fall; the Connecticut warbler does, too, and it makes a suspiciously similar *zeep*.

Under the right conditions, as Libby found, anyone can hear these calls during spring and fall migration. Identifying the species making them, though, can be tough. Many of these calls are short, unremarkable, and very similar across groups of related species; there are the tiny, dry, high-pitched *zeep*s and *tseet*s of warblers and sparrows (which can be less than a twentieth of a second long), the clear liquid *pweer*s of thrushes, the gurgling croaks of cuckoos. Sometimes ornithologists in the decades following Libby's observations got lucky and caught a bird making a nocturnal call they recognized during the day. Sometimes they made educated guesses based on what migrants were passing through an area in the greatest numbers when they heard certain calls. Sometimes they were even able to spot a night-calling bird land as dawn approached and identify it in the growing light.

In the 1940s, the Yale ornithologist Stanley Ball was able to teach himself to distinguish between the night calls of different thrush species well enough to determine when each passed through Quebec's Gaspé Peninsula each fall and what routes they took through the area by sound alone. Ball's work demonstrated the potential of nocturnal flight calls as a tool for studying migration. As remarkable as Ball's efforts were, however, research that relied on the ears of individual observers was hard to scale up or reproduce. In the 1950s, two researchers in Illinois—an ornithologist named Richard Graber and an engineer named Bill Cochran—made the first attempt to record nocturnal flight calls so that they could preserve them and analyze them at their leisure.

The two met thanks to Graber's interest in the phenomenon of "tower kills," flocks of migrating birds being killed en masse by flying into communication towers. Cochran, then in his mid-twenties, was working for the fledgling Illinois television station WCIA and studying electrical engineering part-time. He went out one morning to check on his employer's tower on the outskirts of Urbana and found Graber and his wife there picking up dead birds off the ground. Cochran was intrigued, and soon he was spending his spare time tinkering together gadgets to help with Graber's work at the Illinois Natural History Survey.

Graber's interest in migration soon led him to flight calls, but in a pre-digital world recording nocturnal flight calls wasn't a simple task. For their first attempt in the fall of 1957, Graber and Cochran, wanting to record a full eight-hour night, jury-rigged a tape recorder with a bicycle axle to hold special reels that carried more than *six thousand* feet of tape. To gather and amplify sound from the sky while minimizing sound from the immediate surroundings, they built a wall of straw bales around an enormous parabolic reflector. And creating the recording was only the first hurdle, because they then had to spend *more* than eight hours listening to the whole thing to figure out what they'd captured, sitting through tedious silences and then pausing and rewinding to try to identify any recorded calls. Eager to simplify things, they got a colleague to help them devise a timer that would open and close the circuit of the recorder's reel-turning motor, letting them take a ninety-second sample every ten minutes and compress eight hours of darkness into just over an hour and ten minutes of hopefully representative snippets.

Graber and Cochran kept it up for three years, and in 1960 they published a summary of what they'd learned about migration through Illinois as a result. Their results confirmed the earlier suspicions of ornithologists that big migratory flights depended on favorable winds and the movements of cold fronts. They also showed that daytime bird surveys could provide only an incomplete picture of migration through an area; sometimes after recording a night during which the calls streaming overhead in the dark seemed endless, they would head out for a morning bird-watching session in the woodlands around their recording station and find only a handful of migrants.

Graber and Cochran eventually moved on to other adventures; we'll meet them again later in the book. It would be more than two decades before someone else picked up nocturnal flight call research where they'd left off.

Tuning In to the River of Birds

I realized early that if I wanted the full story of nocturnal flight calls, Bill Evans needed to be number one on my list of people to interview. Googling around for resources related to the subject led inexorably to his website, oldbird.org. There, you can browse a library of flight call recordings organized by species, order a kit to set up your own DIY recording station, and download software to help you analyze what you find. This isn't the first book he's appeared in, but Evans cheerfully recounted his story to me via a phone call in the spring of 2021.

In May 1985, Evans was in his mid-twenties, delivering pizzas in Minneapolis, taking college classes in science and the humanities that weren't really leading toward a degree in anything specific, and going birding every chance he got. One night he camped on a bluff overlooking a river in Afton State Park, hoping to get an early start looking for birds the next morning. He wasn't expecting to discover his life's calling that night. But that's what happened.

Lying awake in his tent, he suddenly realized he was hearing the calls of migrating birds passing overhead in the darkness. In an hour he counted a hundred cuckoos, following each one with his ears as its gurgling trills approached from the south and then dwindled away toward the north. Cuckoos are secretive birds, hard to spot in daylight when they're perched motionless in the treetops, their cinnamon-brown plumage blending in with the branches. But here was an invisible river of them passing above Evans's tent under cover of night. Suddenly, remembering John James Audubon's descriptions of massive flocks of migrating passenger pigeons filling the sky for days at a time, Evans had what he later described as a vision: this was how he could make his mark on the world, recording the sounds of migration for posterity.

He didn't waste any time. The next day he walked into an audiovisual store and asked how he could record "all the sound coming down from the sky at night." Who knows what the clerk made of that, but he recommended that Evans try using just the soundtrack of a hi-fi VCR. Evans scraped together a thousand dollars to buy a VCR, the height of AV technology at the time, and tinkered and experimented until he finally made his first night recordings in the fall of 1986.

Evans never finished his college degree, but eventually he talked his way into a job as a technician at the Cornell Lab of Ornithology's natural sounds library. He spent his days archiving other people's recordings, ate dinner in his rented yurt across the street, and then would return to the lab and stay until midnight, combing the collection for the sounds of nocturnal flight calls. To his disappointment, however, he discovered that they had almost no recordings that could help him identify the night sounds he was obsessed with.

So after only a year and a half, he left, following weather fronts back and forth across the country, recording migration, trying to piece together the identifications of the birds whose calls he captured, and picking up odd jobs to support himself. When he thinks of that time, he told me, he remembers "the sheer enjoyment of sitting up on a hilltop, cracking a beer, and listening to those flights." It made him feel in tune with the whole planet: no matter where you are, "if you can tune in to this river of birds going over in migration, it connects you, your mind expands."

After another stint at Cornell, this time as a research associate working on developing some of the first software to help analyze flight calls, Evans faced the realization that there just wasn't enough interest out there in migrating birds' nocturnal sounds for him to get any traction for the kind of work he wanted to do. So in 1998 he founded a nonprofit he dubbed Old Bird to create that interest base himself.

The name Old Bird, Evans told me, came from his idea that the flight call data he was collecting was going to be the "old" bird information of the future, like Audubon's account of the migrating passenger pigeons. "I remember feeling very depressed, thinking that there was a lot of environmental destruction going on and things were only going to get worse. The way I got out of that and turned it into a positive vision was thinking that at least I could leave a record of what's here now for the future." Through Old Bird, Evans set about developing and selling relatively inexpensive DIY kits for recording nocturnal flight calls at home and making identification resources and software for analyzing the calls available to the public.

In collaboration with a birder named Michael O'Brien, in 2002 Evans released a CD-ROM guide to flight calls of birds of eastern North America. The project took about a decade to complete. O'Brien, who connected with Evans after buying a cassette tape of thrush calls that Evans had been selling through ads in birding journals, moved to the migration hot spot of Cape May, New Jersey, to record flight calls there, while Evans continued traveling to record in other areas. They didn't want to release the guide until they had pinned down the calls of every possible species, and discriminating between the very similar little *zeeps* of some warblers and sparrows required a lot of puzzling over the characteristics of individual recordings and ruling out possibilities one by one.

"You know when you're doing a jigsaw puzzle, how hard it is at the beginning, but then once you get the edge together and you have a little more structure, you're on your way?" said Evans. "And as the pieces start to fall, it goes together very quickly at the end." Today the material from the guide is available for free through www.oldbird.org.

"The phenomenon right now that's exciting is all the other people getting into it, in Europe, North America, just so many people contributing," said Evans. "I remember early on, I thought I sort of had a handle on these calls and what was going on. Now every year there are just so many unknown and weird things that other people are finding that it makes me realize how much more there is to know."

Rise of the Machines

Just south of Missoula, Montana, the Bitterroot Valley is a long, narrow stretch of land that lies between the Bitterroot Range to the west and the Sapphire Mountains to the east. The Bitterroot River, lined with pine trees, winds between cattle ranches along the valley floor. In 2018 and 2019, this quiet corner of the West was home to the largest-scale attempt to record nocturnal flight calls to date.

Debbie Leick and Kate Stone work for MPG Ranch, a private seventeen-thousand-acre research and conservation property in the valley. In 2012, they attended an ornithology conference in Vancouver, British Columbia. There, they heard a series of talks on acoustic monitoring and decided they had to try it out for themselves.

With few resources out there for those just getting into nocturnal flight calls, the project got off to a slow start. "I had no idea what I was doing. The first couple of weeks were pretty rough," said Leick when I first spoke to her and Stone in the spring of 2021. "We started with three microphones, [one each] at a low, mid, and high elevation. They corresponded with our banding stations, and the intent was to try to correlate the data from the recorders with the banding data."

But in the following years they added more recording sites around the valley, reaching out to homeowners, local schools, and the Bitterroot National Forest to ask them to host recording stations. Soon the project had grown from three microphones to six to a couple dozen. One, located at a fire lookout operated by the Forest Service at the top of a peak in the Bitterroot Range, required an eight-mile round-trip hike to install the equipment.

In the fall of 2018 and 2019, they operated fifty-two recording sites, which as far as Leick and Stone know makes their "Bitterroot Array" the biggest-ever nocturnal flight call recording project. Leick and Stone are still working on analyzing the massive amounts of data—more than 100,000 hours—that they collected over those two seasons. Some of the birds that popped up most frequently in their recordings, like Swainson's thrushes, Wilson's warblers, and upland sandpipers, are species they rarely see during daylight. The recordings are also giving Leick and Stone a new sense of how migrating birds move through the valley, with more calls recorded at topographic pinch points where the valley narrows, and more at lower elevations than at higher elevations. They even identified the nocturnal flight call of a gray-cheeked thrush, a species never before observed in Montana.

By the time I contacted them, they'd scaled back down to three sites, which they intend to keep going as a long-term monitoring project. After first speaking to Stone and Leick via video call, I asked if I could visit the ranch in person, and they obliged. So on a crisp morning in August 2021, Debbie Leick led me to one of the three ongoing monitoring sites, overlooking the pines along the Bitterroot River.

A tiny microphone built by Bill Evans, housed in a plastic two-and-a-half-gallon bucket atop a sixteen-foot wooden pole, gathered the sound of birds passing overhead. Sticking out next to the bucket at the top of the pole was a perch that Leick explained was for juvenile great horned owls waiting to be fed by their parents. If it weren't for the perch—and some spikes added to the bucket itself as a deterrent—they would sit directly on the microphone and "beg and beg and beg and beg the whole night and basically trash your recording," she said. The rest of the recording equipment was housed in a sturdy tote at the base of the pole, its latches reinforced with zip ties to keep out curious feral horses, topped by a solar panel that powered the whole thing.

Leick didn't know anything about electrical components when she started the project. "Now I feel like if I had to, I could probably wire a house," she said as she checked on the equipment in the tote. If she can ever get funding, she has daydreams about how to make even better use of recording stations like this, like using thermal cameras to match call recordings with infrared images of birds passing overhead, or putting microphone backpacks directly onto migrating birds to gauge how frequently individuals call.

Leick checked the batteries on the recording unit and then pulled out its memory card, carrying it back to the tailgate of her vehicle, where she plugged it into her laptop to make sure everything was working correctly. "Okay, it's loading. Come on, please, please let there be—oh, it recorded. Awesome." There on the screen was a spectrogram representing hours and hours of sound. On that particular night, the recorder had mostly captured the sound of the wind and a few insects, but that, too, can be valuable data if you want to know when birds are and are not most likely to be passing through.

Bill Evans's guide to flight calls covered only eastern North America, leaving some gaps in their ability to identify some of the calls they recorded. At the height of their project, Leick and Stone resorted to tactics like placing captured birds in a mobile recording booth after they'd been banded to see if they could coax them into producing their nocturnal calls and leaving microphones out in specific habitats that they knew were frequented by only certain species. But Leick and

Stone haven't been alone in their efforts to tally and identify the calls of birds passing over their valley at night. They've had help from a source that would probably have seemed like science fiction to Bill Cochran and Richard Graber: artificial intelligence.

The first big computer-driven advances in analyzing nocturnal flight calls came in the early to mid-1990s. One was the introduction of user-friendly software for converting recordings of birdcalls into easy-to-analyze spectrograms, graphs that depict the shapes of individual calls with time along the horizontal axis, pitch (low to high) along the vertical axis, and loudness indicated by color. Another was the development of some of the first "automated detectors," which Bill Evans and the computer scientist Harold Mills worked on together at the Cornell Lab of Ornithology. It still wasn't practical at that point to record sound directly onto computer hard drives, so they would play recordings from a VHS tape into a computer, which would listen for individual calls and clip them out as small sound files. This was an attempt to solve the problem Graber and Cochran had run up against in the 1950s: listening to what can be hours of silence to pick out and identify bursts of birdcalls is much less tedious for computers than it is for humans.

Mills left Cornell in 2005, but he's still working on the same problems today. His current project is called Vesper, a piece of open-source software for helping people organize and analyze large volumes of nocturnal flight call data, like those collected by the Bitterroot Array. Mills has worked with Debbie Leick at MPG Ranch to develop new and improved detectors using her data.

The ultimate aspiration for computer scientists working on nocturnal flight calls is to teach computers not only to detect the difference between birdcalls and silence but to automatically identify the species producing the calls—in the jargon of the field, to have "classifiers" as well as detectors. One project working toward this is BirdVox, a collaboration between researchers at the Cornell Lab of Ornithology and NYU's Music and Audio Research Laboratory. Since

2016, the BirdVox team has been working on software that they hope will eventually be able to classify birds at multiple levels, telling researchers, for example, whether the bird they've recorded is a sparrow or a warbler even in situations where it's not possible to nail down the exact species.

Modern detectors and classifiers make use of what are called neural networks, computing systems loosely inspired by the structure of the human brain. You interact with a neural network each time you mark an email in your in-box as spam or use the predictive text feature while tapping out a text message. All input into a computer—whether a spammy email, a photo, or the nocturnal flight call of a blackpoll warbler—is converted into an array of ones and zeros, and a neural network sorts through all those ones and zeros in search of patterns. Eventually it learns that a particular pattern of numbers is associated with, say, the call of a specific sparrow species. The next time it finds a pattern that looks similar, the neural network can tell you that that pattern might be that kind of sparrow, too, and estimate how confident it is in that identification.

A neural network typically needs a *lot* of data to chew through before it gets any good at picking out patterns, because it can't draw from past experiences the way a human can; a bird-watcher can travel to a new part of the world and start to learn the calls of local birds after hearing just a few examples, but a neural network might need tens of thousands of individual recordings before it can make sense of what it's "hearing." And while there's no shortage of spam emails for neural networks to practice on, massive flight call data sets can be harder to come by, which is why Mills was eager to try using the data from MPG Ranch to train new detectors.

Vesper and BirdVox are both open-source, meaning anyone can download and play around with them. So far, though, computers can't beat the ears of human experts at identifying all those *zeep*s and *pweer*s; background noise, variation in exactly how calls are recorded, and other factors that our brains can easily adjust for still trip up artificial neural networks.

Ninety Thousand Thrushes

There's no one perfect method for learning about bird migration, and nocturnal flight call recordings are no exception. For one thing, not all birds call all the time. Some might call only at certain times during the night, while others remain almost completely silent, and recordings can't tell you anything about birds that keep quiet. A lot of the studies that have been done on flight call data in recent years have been what you might call "proof of concept" studies—for example, comparing data from flight call monitoring with data on what birds are being captured at bird banding stations at the same time, to make sure that they match up well.

But for sheer numbers of birds identified, flight call studies are hard to beat. "If you put out a hundred [radio transmitter] tags on a species, and you get a couple detections, it tells you one thing," MPG Ranch's Kate Stone told me, "but you know, we're picking up ninety thousand Swainson's thrushes in a season. That's pretty incredible."

In the past two decades, scientists have used nocturnal flight calls to study migration patterns over the Gulf of Mexico, across Lake Erie, and along the coast of Rhode Island. Flight call recordings have revealed the migration routes that seabirds called scoters follow through the British Isles, and that night-migrating birds are attracted to artificial lights at ground level (not only on tall skyscrapers and communication towers, as past studies had shown).

The introduction of low-cost recording equipment, open-source software, and better identification resources has also brought nocturnal flight call recording within the reach of hobbyists. As of the spring of 2021, a Facebook group for flight call enthusiasts in the United States and Canada (which Bill Evans is active in, helping others identify the species they've recorded) had almost two thousand members. And though work on the flight calls of European birds has not advanced as far, another Facebook group based in Europe (where devotees refer to the nocturnal migration phenomenon by the nickname nocmig) had about eleven hundred members.

One member of that group, Martin Minařík, moved to England from the Czech Republic in 2017 for a postdoctoral research position in developmental biology. He discovered flight call recording during the coronavirus pandemic while recovering from "long COVID"—the lingering, sometimes debilitating fatigue and other symptoms experienced by a portion of people who get sick with COVID-19. Already a bird-watcher, he heard the high-pitched calls of a flock of common sandpipers passing overhead while on a short walk with his wife after dark in August 2020. Soon he bought a cheap microphone and, unable to go on the lengthy bird-watching outings he'd loved before his illness, found excitement each time he recorded another species passing over his home. His health is gradually improving, but he told me via a Facebook message that nocturnal flight call recording is "definitely a hobby that's gonna stay."

The flight call researchers I spoke to envision a time when every bird banding station will also have a microphone array for monitoring nocturnal flight calls—a time when ornithologists, volunteers, and hobbyists like Minařík will be able to run a ten-hour recording through a piece of software that instantly tells them what bird species were detected and in what numbers, and when the data collected via their microphones will be one more crucial tool for monitoring and conserving migratory bird populations. We're not there yet. But in a few more years, we may be.

Moonlight Magic

Hearing migrating birds at night is one thing; *seeing* them, flying through the darkness high overhead, is another. But on the night of October 19, 1880, the ornithologist William Earl Dodge Scott, taking a tour of the astronomy department at Princeton University (then called the College of New Jersey at Princeton) with some friends, accidentally discovered a way to do just that. When Scott and his companions were shown the moon, just a few days past full, through a telescope, Scott was astounded by what he saw: not only the craters and dark "seas" of the surface, but also silhouette after silhouette of small birds, flying across the moon's bright face.

"My attention was at once arrested by numbers of small birds more or less plainly seen passing across the field of observation," he wrote when he later described the experience for an ornithological journal. "They were in many cases very clearly defined against the bright background; the movements of the wings were plainly to be seen, as well as the entire action of flight. In the same way the shape of the head and the tail were conspicuous when the bird was well focused."

Scott could see them so well that he could pick out warblers, finches, and woodpeckers by their shapes. Most were flying toward the southeast. He was, he quickly realized, witnessing fall migration in action.

He didn't describe what his tour guide and companions thought as he brought the evening to a grinding halt to start collecting data, but Scott began taking notes, recording four to five birds passing through his field of view every minute. Back at his desk after the experience, he tried to calculate how high they must have been flying, based on how far they would have needed to be from the telescope to appear in focus. His estimate of one to four miles was probably too high (modern data shows that birds migrating over the eastern United States are usually less than half a mile up), but it was a valiant attempt.

In the ensuing years, a few other ornithologists would follow Scott's lead and use "moon watching" observations to calculate approximately how high passing birds were flying. But not until the 1940s would a clash between competing migration theories bring moon watching to its full potential.

Ornithologists had speculated that at least some birds migrated across the open waters of the Gulf of Mexico to reach Central and South America as far back as 1905. Birds had been directly observed arriving and departing along the Gulf Coast, and many species that bred in North America and wintered in the Neotropics were almost completely unknown in eastern Mexico, where they'd need to travel if they were going around the Gulf instead of across.

In the early 1940s, however, the biologist George C. Williams developed an alternate notion. He believed that it was impossible that birds were routinely passing back and forth across more than five hundred miles of open ocean, and argued that migrants passing through eastern Mexico simply stuck to a narrow route along the coast and, most of the time, flew too high to be observed.

In Louisiana, a young ornithologist named George Lowery—the very same George Lowery who would later hand the bat biologist Walter Dalquest some Japanese mist nets to try out on his expedition to Mexico—was frustrated by Williams's assertions. Attending college in Baton Rouge in the 1930s, Lowery would have had ample opportunities to see firsthand the amazing "fallouts" that sometimes happened there, where exhausted songbirds reaching land after their marathon trans-Gulf flight would fill the forests. By the 1940s, he'd finished his master's degree at Louisiana State University and stayed on to found the LSU Museum of Zoology, and he was determined to prove Williams wrong. To do so, he needed a method to measure the amount of migration taking place at different times and different areas, and it had to work at night, when most migration occurs.

Luckily, Lowery remembered reading some old papers published around the turn of the twentieth century about a quirky technique that involved observing birds flying across the disk of the full moon. Lowery turned to William Rense, a bird-loving astronomer then teaching at LSU, who developed a series of equations to turn raw data on the number and direction of birds passing across the moon in a given unit of time into an estimate of the total number of birds crossing an imaginary one-mile line on Earth's surface every hour. Lowery eventually dubbed this the "migration traffic rate," and with this method he could directly compare the amount of migration at sea in the Gulf of Mexico with that over land in eastern Mexico and anywhere else.

Or so he hoped. On April 29, 1945, Lowery set sail from New Orleans on the SS *Bertha Brøvig*, bound for the Yucatán. In her excellent 2006 book, *Songbird Journeys*, the author and ornithologist Miyoko Chu described what happened next:

> That night, he set up his telescope on deck and aimed the lens at the moon. As Lowery peered through the telescope's eyepiece, he hoped to see the silhouettes of birds crossing in front of the moon, but no matter how he steadied himself against the rocking of the boat, the moon kept bobbing in and out of view. With the telescope useless at his side, Lowery could only have gazed up in the darkness, knowing that thousands of birds could be passing between him and the moon, but that there was no way to see them.

But Lowery didn't give up easily. He spent the rest of his time at sea recording every passing land bird he spotted from the ship, tallying twenty-one species over the five-day round-trip. And while the ship was docked and still in the Yucatán, he tried moon watching a second time, pointing his telescope at the moon rising over the Gulf in the wee hours of the morning. In about forty-five minutes, he tallied twelve birds. Run through the equations of his astronomer friend Rense, however, that worked out to more than thirty-seven hundred birds passing over a one-mile line extending out into the Gulf every hour—streaming past over the open water, in defiance of George Williams and anyone else who doubted that the amazing phenomenon of trans-Gulf migration was real.

Lowery knew he was onto something, and his ambitions for his new moon-watching technique went much, much further than a night or two in Mexico. "Flight studies by means of the moon were much more than a way of finding out whether any migrants cross the Gulf," as a colleague of his later wrote in *Audubon Magazine*; "they would permit us to peer into the very heart of age-old basic mysteries of mass migration." Lowery would devote the next two decades of his life to making his dreams for moon watching a reality.

Telescopes Swinging into Action

Bob Newman might have seemed like an unlikely partner for George Lowery when he arrived at LSU as a graduate student in 1945, just as Lowery was developing his new moon-watching technique with William Rense.

Lowery—or "Doc," as he was known to everyone who worked with him—was a naturalist of the old school, taking an interest in mammals as well as birds and sending graduate students on regular expeditions to Central and South America to collect specimens to taxidermy for the museum's collection. His students remember him as someone who would go home to eat dinner but then come back to put in another hour or two of work in the evening. Even a diabetes diagnosis couldn't slow him down, and new grad students were advised to keep hard candies in their desks in the museum in case Lowery came through in the afternoon slurring his words from hypoglycemia. (Lowery wasn't *all* business, though: he liked to begin each advanced ornithology class during football season with a fifteen-minute recap of the most recent LSU football game.)

Bob Newman, on the other hand, quickly developed a reputation as a jokester, always telling little riddles with a twinkle in his eye. While Lowery would have been at home among the gentleman ornithologists of the previous century, Newman had an eye on the future, teaching himself statistics before it was a common skill for zoologists and making it a personal mission to keep up with the most current research on bird migration from around the world. Rather than clashing over their contrasting approaches to their work, however, the two men forged a partnership that would last until Lowery's death in 1978. Like Lowery, Newman never left LSU, staying on to work at the newly formed zoology museum after he finished his degree.

Before Lowery set sail for the Yucatán in 1945, he recruited thirty colleagues across the central and southern United States to make their own moon-watching observations of spring migration and send him the data. Newman helped him analyze the results, and what they found intrigued them. The data consistently showed that the number of migrating birds peaked in the middle of the night before gradually dwindling away until dawn. Ornithologists had once supposed that night-migrating birds flew *all* night, or perhaps were staying up late or getting up early to extend flights that happened primarily in daylight; the rate of nocturnal flight calls tended to peak just before sunrise.* Moon-watching data, on the other hand, suggested that birds "must get up in the middle of the night just for the sake of making three or four hours' progress toward their destinations," as Newman later put it in an article he wrote for *Audubon Magazine*.

But this only made Lowery and Newman hungry to learn more. Did the same pattern hold true across broader geographic scales? What about in the fall, as opposed to the spring? What about the effects of wind and rain and mountain ranges and all the other obstacles birds might encounter? They needed more data. And so the two men hatched a plan to collect a snapshot of migration activity across the entire continent during a single week in October.

Nothing like it had ever been attempted before. Today, when we're all used to the convenience of the internet, it's almost hard to imagine what must have gone into coordinating such a project in the 1950s. For Lowery and Newman, the only means of organizing this massive undertaking was paper—mountains of it. Newman described their recruitment efforts in his doctoral dissertation:

> The Museum began an all-out campaign to recreate a continentwide network of observation stations and to build up an accumulation of autumnal data. . . .

---

* Modern research suggests that the time when flight calls peak doesn't really reflect the time when overall migration peaks; birds are just easier to hear when they're descending before dawn, and may call more frequently then as well.

Announcements were inserted in ornithological and astronomical journals. Personal letters were written to hundreds of potential participants, stressing the special contribution that each might make to the coming effort. This correspondence involved more than 12 hours of work a day, but it brought me many pleasant associations with people all over the country and was perhaps a major factor in the fantastic success of the campaign.

In the end they recruited twenty-five hundred volunteers in 325 locations from Canada to Panama.

Lowery and Newman also wrote up a how-to pamphlet for observers to make sure the data collected was as reliable as possible, complete with advice for how to arrange pillows and an adjustable lawn chair to stay comfortable while gazing at the moon for long periods of time. Moon watchers described some nights where tiny black silhouettes would flash across the moon as fast as observers could record them, but on other nights the work could be incredibly tedious, with perhaps one bird passing in half an hour of staring into the full moon's headache-inducing glare. Lowery and Newman even recommended that their volunteers wear a patch over the eye not peering into the telescope.

"When the autumn of 1952 once again brought the birds southward, a wave of enthusiasm for moon-watching swept the nation," Newman wrote in his dissertation; he'd majored in English as an undergraduate, and his flair for language regularly showed itself in his scientific writing. "Telescopes swung into operation at more than 300 localities as people by the thousands took up the new form of bird study. By the end of the season, reports had been received from every state in the United States and all but one of the provinces of Canada."

But as observations flooded in, paper data sheets mailed from every part of the country stacking up on their desks, Lowery and Newman realized they had another problem: what to do with all that data.

This was an era before computers made sophisticated statistical analysis relatively simple and before graduate students in zoology were expected to take courses like calculus. They needed to calculate the average direction and volume of bird migration from hundreds of locations. They needed to look up data on temperature and wind and the movement of cold fronts for each of those locations, and check for patterns in how migration related to weather across the entire continent. They needed to depict all of this on easy-to-interpret maps, which they would have to draw by hand. "As reports and inquiries poured in," wrote Newman, "our facilities for handling them proved slender by comparison with the work that needed to be done."

Lowery and Newman took fourteen long years to complete and publish their analysis. In 1966, their magnum opus finally appeared in the ornithological journal *The Auk*. They titled it "A Continentwide View of Bird Migration on Four Nights in October."

Their painstakingly crafted maps used arrows colored red, black, or white to indicate the volume as well as the direction of migration at locations across the continent on the nights of October 1–4, 1952. The patterns that Lowery and Newman revealed are familiar to modern bird-watchers eagerly watching the weather forecast in spring and fall to try to predict when the next big wave of migrants will reach their area. In the Midwest and East, massive migratory flights spanned multiple states, with reports from many locations providing remarkably consistent data on how many birds were passing over and what precise direction they were heading. Favorable winds and the movement of cold fronts both clearly affected migration patterns across the continent, with the heaviest migration occurring in front of advancing cold fronts. And migration traffic was consistently far higher in the eastern half of the continent than west of the Rockies. Look at a map and imagine a straight line from the northern edge of South America to anywhere in Canada, and you can immediately see that birds heading to North America's northern reaches will mostly bypass the western United States.

The work was a blockbuster, influencing studies of migration for generations. It was also the last moon-watching paper that Lowery and Newman would ever publish. They planned to continue their efforts, adapting their telescope methods for use scanning the skies in daylight and applying for funding for a new study of migrating birds arriving on the Gulf Coast. But the technique they'd developed was about to be made mostly obsolete by one of their own students and the rise of a new technology: radar.

But we might not have heard the last of moon watching quite yet.

In 2020, a group of scientists from the University of Oklahoma, inspired by a conference presentation on Lowery's work, published a paper describing their design for a new, automated moon-watching system they've dubbed LunAero. To eliminate the need for human observers to stare at the moon for hours on end, LunAero uses a motorized spotting scope mount to follow the moon across the sky over the course of the night and a camera to record video. Its creators hope to make it simple and inexpensive enough for hobbyists, similar to the work Bill Evans did to bring flight call recording to the masses. Using video recordings allows for the collection of far more detailed and precise data, and the scientists behind LunAero hope to use it to study phenomena like social behavior, the question of when migrating birds fly singly versus in groups.

I don't have a LunAero setup, but I still couldn't resist giving moon watching a try myself. The April 2021 full moon came just a week and a half shy of the peak of spring migration in my area, and I arranged to borrow a small telescope to set up in my driveway.

My husband and I watched a couple episodes of TV while I waited for the moon to climb far enough into the sky above the mountains on the horizon for a clear view. When I went outside, the warm April day had been replaced by a night cool enough that I needed a jacket. It took only a few moments to locate the moon's bright disk in the telescope eyepiece. Listening to frogs calling in the distance while my eyes adjusted, I admired the dark seas of ancient lava and the round pockmark of Tycho Crater, whose diameter moon watchers in the 1950s used as a quick way to estimate the relative size of the bird silhouettes they tallied. (One of my favorite bits of Bob Newman's writing was his description of Tycho Crater as "the navel on an immense floodlit plaster model of an orange" in the article he wrote about moon watching for *Audubon Magazine*.)

It was a beautiful sight, but as the minutes ticked by, not only did I not see anything resembling a bird, but my back and neck started to ache from hunching over the eyepiece. I switched from standing to sitting on a chair to sitting on a stool and adjusted the length of the telescope tripod's legs, but I couldn't find the perfect position. I suddenly understood why Lowery and Newman's instruction manual for volunteers had included exhortations about the importance of a comfortable lawn chair. I kept having to relocate farther and farther down my driveway to keep my view clear as the rising moon became entangled with the branches of a tree. Whenever I lifted my face from the eyepiece, my vision swam.

Before I gave up, I did glimpse one small black speck streak across the moon's surface, so fast it was gone before I registered it. As a very inexperienced moon watcher, I'm not completely certain that it wasn't an insect (or, on the opposite end of the distance spectrum, a satellite—a distraction that Lowery and Newman wouldn't have needed to worry about). But maybe it was an oriole or a tanager or—my favorite of the migratory songbirds where I live—a turquoise-hued lazuli bunting, just arrived from Mexico. In the coming weeks I would see all those birds and more as returning migrants filled the nearby woods once again.

Three
Chasing Angels

S omewhere off the coast of China in the Yellow Sea, Irven Buss was watching blips on a radar screen.

It was October 19, 1945. As a fighter director officer for the U.S. Navy, Buss had been charged with interpreting information from the ship's radar operators and using it to direct the movement of fighter planes in the air. It was sophisticated work that required extensive training. Military radar operators had figured out in the preceding years that the mysterious shadows they sometimes picked up were flocks of birds in flight, and Buss was aware of several incidents in the spring of 1945 where navy vessels picked up radar targets moving at various speeds and were able to visually confirm that they were flocks of birds. "That these birds were not positively identified," Buss drily put it when he later wrote about that night's events, "is largely due to the fact that few radar operators are ornithologists."

An ornithologist, however, is exactly what Buss was.

Before joining the navy, Buss had been the chief of wildlife research for the Wisconsin Conservation Department (now the Wisconsin Department of Natural Resources), where he studied sandpipers and pheasants. As a fighter director officer, he was now in an ideal position to remedy the ornithological shortcomings of past radar operators, and he was determined not to let this opportunity pass him by. "Continued vigilance for the mysterious signals was maintained in the hope that eventually such targets would be detected on the ship's radar," he wrote in his 1946 paper for *The Auk*.

On that night in October, he finally got his chance. Between 8:11 p.m. and 5:11 a.m., his ship's radar system tracked five of his "mysterious signals" traveling low to the water in a southward direction, at speeds between twenty-six and thirty-one knots. Buss spent a sleepless night noting down the characteristics of the blips as they blinked across the screen, certain they could only represent passing birds.

The break of dawn proved him right. As the light strengthened, the movement of beating wings slowly became visible over the gray sea. The ship was surrounded in every direction by flock after flock of migrating ducks, each flock containing as many as a hundred birds, all flying two hundred feet or less above the surface of the sea and heading south.

"A direct and efficient communication system between the radar station below decks and a visual lookout station topside," Buss wrote, "made it possible to check radar targets by ocular observation."

I like to imagine Buss twisting the arms of his non-ornithological shipmates to help him collect his data, quickly teaching them how to spot the difference between the green heads of mallards and the brown heads of pintail ducks, all of them calling back and forth to compare visual observations with the signals continuing to be picked up by radar. Japan had formally surrendered to Allied forces just six weeks before, and it must have been fun, even celebratory, to use the ship's radar system for something other than war.

Some flocks that they could follow visually were too small for the radar system to track. Most of the larger, easier-to-track flocks stayed too far from the ship for the sailors to get a good look. However, Buss and his impromptu research assistants managed twice to track a flock on radar and measure its speed while also following it visually to confirm the number and species of birds. One was a flock of about fifty pintails, the other a mixed flock of fifty ducks that were mostly mallards. The characteristics of the radar signals and their overall speeds matched those Buss had tracked overnight. In his paper describing his observations, Buss noted that these were much slower than flight speeds previously measured for mallards and pintails pursued by hunters and speculated that they must travel more slowly while migrating.

Buss was likely the first person to intentionally use radar to gather data about migrating birds and draw some conclusions about their behavior. But by the time of his morning observing ducks on the Yellow Sea, years of work had already gone into documenting that birds showed up on radar at all.

In 1935, British scientists had demonstrated that a shortwave radio transmitter owned by the BBC could be used to detect bombers up to eight miles away, and by the time World War II officially broke out, radar stations had been built along the south coast of England to detect incoming planes. However, it quickly became clear that aircraft weren't the only objects being detected by radar systems. Planes typically plodded along regular paths, their speed, direction, and elevation more or less constant, but radar operators also reported mysterious echoes "moving at between 5 and 80 miles per hour, often against or across the wind. Some would fade within seconds; others persisted to form prolonged tracks, some exceeding 60,000 yards [approximately 34 miles]. Many would annoyingly wax and wane along a trajectory, making visual detection and verification difficult." On at least one occasion, RAF Fighter Command sounded a red alert and British fighter planes scrambled to intercept mysterious radar signals crossing the English Channel, only to find nothing there. Sometimes the strange signals could be traced to floating wreckage or other solid objects, but most remained unidentified, and exasperated radar operators began to refer to them as "angels."

It was a young science teacher named David Lack who ultimately uncovered the earthly explanation behind the angels. Lack had always been interested in ornithology; in 1938, he took a year off from his teaching job to travel to the Galápagos Islands to study Darwin's famous finches. Returning to England just as war was breaking out, Lack first joined the pacifist movement, but during the London Blitz in the fall of 1940 he was "so put off by the pacifists' earnest attitudes and so excited by the flashes and bangs" that at the age of thirty he decided to join the war effort instead.

Instead of becoming a military officer, Lack soon found himself being interviewed for a mysterious civilian job that would supposedly make use of his background in the sciences. In a memoir fragment published after his death, Lack (who apparently had a sense of humor) recounted his conversation with the interviewer as follows:

"As a biologist, you will, of course, have learned a lot of physics."
"I am afraid not."
"Well, I expect your maths is of a high standard."
"I am afraid not."
"Anyway, you will obviously be good with your hands."
"I am afraid not."
Then, very doubtfully, "I fear this job will often entail going out in the wet and cold in the dark. Do you mind?"
"Not at all."

Lack was hired, and he shortly began his training on the use of the newly developed radar systems.

After a stint in the Orkneys—a remote group of islands off the northern coast of Scotland where Lack spent his days off bird-watching, squinting at puffins and skuas along the islands' windy, rocky coasts—and an "exhausting" six months as an assistant to a lieutenant colonel who was an expert in atmospheric physics, Lack was touring radar installations up and down the British coast when he was reintroduced to an entomologist named George Varley. Varley had been a contemporary of Lack's at Cambridge and had fallen into the same line of work, and it was he who introduced Lack to the "spurious echoes" plaguing radar operators.

As an amateur ornithologist, Lack knew enough about the behavior of birds in flight to suspect that they might be responsible, but proving it took him and his colleagues time. At one point, an officer named Ramsay tied a dead gull to a balloon and sent it aloft into the path of a radar station to prove that birds could, indeed, create radar signals. It was Varley who finally obtained the first concrete proof that birds in flight (as opposed to dead birds hanging from balloons!) could produce radar echoes when, in September 1941, he used a powerful telescope to confirm that the source of a signal being tracked offshore in real time was a flock of large white seabirds called gannets.

Lack and Varley's work confirmed that the airspeeds of "angels" and their tendency to fly with or against the wind (rather than perpendicular to it) were typical of bird behavior, and they continued to collect observations throughout the war of radar angels that were visually confirmed as birds. At first only large birds such as geese and gannets showed up on radar, but as radar equipment became more powerful, it began to detect flocks of smaller birds such as starlings as well.

Lack and Varley's findings were not immediately and universally accepted. "At one meeting," Lack later wrote, "after the physicists had again gravely explained that clouds of ions must be responsible, Varley with equal gravity accepted their view, provided that the ions were wrapped in feathers." Even other ornithologists weren't all so sure at first; not all scientists at the time believed that seabirds were in the habit of flying at night, when many of the angels were detected.

But as the end of the war neared, Lack and his colleagues began to obtain permission to publicly publish their findings on birds appearing as radar signals, which is presumably how Buss learned about the phenomenon. In a 1945 letter to the journal *Nature*, Lack and Varley summarized secret reports produced earlier in the war describing the radar echoes produced by flocks of seabirds. Other publications suggested that the Germans, too, had spent part of the war chasing down spurious radar signals produced by birds.

After World War II ended, the ornithologists who were the first people to study birds via radar largely moved on. Irven Buss switched gears entirely and went on to become a well-regarded expert in the management of elephant populations in Uganda. Lack continued to keep tabs on radar observations of birds in Europe, but today he's mostly remembered for his book about his work on Darwin's finches, which earned him the nickname "the father of evolutionary ecology." (He didn't forget his wartime colleagues, however; Varley was the best man at his wedding in 1949.) Radar might have remained mostly an interesting footnote in the history of ornithology if not for the realization that the technology had applications in a third field, beyond the interests of the military and bird scientists: weather forecasting.

Starry Skies Full of Birds

I'm a bird person, not an engineer, and it took me some time to wrap my head around how radar actually detects birds (or rainstorms, or anything else). At one point I resorted to posting my more technical questions on Twitter, hoping one of the scientists who followed me there might be able to help me out. That's how I first met Phil Stepanian; after he patiently answered a series of questions via tweet, I eventually called him up to bombard him with even more.

Stepanian, a professor in the department of Civil and Environmental Engineering and Earth Sciences at the University of Notre Dame, studied meteorology and engineering before discovering radar ornithology as a graduate student. "It's kind of embarrassing, but until I started grad school, it never even occurred to me that people, like, studied birds scientifically," he told me. "I knew that people did bird-watching with binoculars and put up bird feeders, but it never occurred to me that you might get a PhD in ornithology and that there were scientists who studied birds." But when he started his own PhD in atmospheric science, his advisor at the University of Oklahoma was collaborating with ornithologists to study birds using cutting-edge radar. "The deeper I got down that rabbit hole, the more obsessed I got with it, to the point that now I really only do flying animal work."

"Radar" stands for radio detection and ranging, and radar equipment works by beaming out radio waves from specialized antennas. As the waves bounce invisibly off objects in their path, some of their energy is returned to the radar equipment. The amount of energy returned, the angle at which it comes in, and how long it takes to come back reveal the size and distance of any object the waves ran up against, from a raindrop to a fighter plane.

The basic properties of radio waves and the way they reflected off objects were demonstrated by the German physicist Heinrich Hertz in the 1880s, and the first device to aid ship navigation by detecting obstacles via the reflection of radio waves was patented in 1904. However, there was no real market for the technology until the 1930s, when the first long-range military bomber planes were developed during the buildup to World War II. The United States, Great Britain, Germany, France, the Soviet Union, Italy, the Netherlands, and Japan all developed their own military radar systems at around the same time, with varying degrees of success, but it took time to truly recognize the technology's potential. Although radar detected the Japanese planes approaching Pearl Harbor, for instance, the importance of the signals was not realized until after the attack began.

Birds weren't the only non-airplane signals showing up on military radars during World War II. Heavy precipitation and even, under the right conditions, the boundaries between two air masses could reflect radio waves and show up in radar readings. In 1946, the U.S. Navy donated twenty-five surplus radar sets to the U.S. Weather Bureau, now known as the National Weather Service, and the era of weather radar was born.

After early weather radar installations proved immensely useful for tracking major hurricanes that hit the East Coast in 1954 and 1955, the Weather Bureau persuaded Congress to fund what became the WSR-57, the first modern weather radar network. (The designation

57 came from the year that the network's design was completed.) The WSR-57 network began with thirty-one new radar sets installed along the Atlantic and Gulf coasts to detect hurricanes and in the Midwest to detect severe local storms.

Modern weather radar is different from the military radars of World War II because it has a different goal. Radars like the ones Buss and Lack worked with had relatively high resolution, showing airplanes and even single birds as individual blips on a screen, but operated over comparatively short distances. Weather radar is a "system that has much longer ranges, so it can look out a hundred miles, but doesn't have that fine-scale detail," said Stepanian. "So you wouldn't be able to see an individual bird, but you would see big flocks or groups of birds, the same way you can't see a specific raindrop but you see a cloud or a storm."

Feathers alone don't show up well on radar; a bird's radar reflectivity comes from the water contained within its blood and muscles. (Swarms of insects can also show up on radar, in their case thanks to chitinous exoskeletons.) On weather radar, migrating birds often appear as a sort of big donut-shaped blob, which Stepanian explained is the result of birds filling a radar station's entire cone-shaped field of view as they take off en masse.

Computer programs remove these blobs of birds from the radar images you see on the Weather Channel. This is about more than just not confusing television viewers; weather forecasts are often based on using current radar data to project what will happen in the near future, and bird data has to be filtered out so it's not mistaken for, say, a massive rainstorm. "Meteorologists have sort of developed a hatred of birds, because they're messing up all of the algorithms that they worked so hard to develop," joked Stepanian.

There are many things weather radar can't tell us about birds. Unlike nocturnal flight call monitoring, it can't tell us what specific bird species are passing overhead, a fact that has caused headaches for more than one radar ornithologist who's delivered a talk about their cutting-edge research at a scientific conference, only to have a confused audience member raise their hand at the end and ask, "But what species do you study, exactly?" Nor can it tell us exactly how many birds are in a blob, although ornithologists have come up with ways to estimate this.

But we can still learn a lot from radar about large-scale patterns of bird migration. "I would say it tells us something about relative abundance, distribution, and seasonality of migrating birds," said Stepanian. "So it can tell us that there are more birds in this area than that area, which is useful if we want to know where the birds are taking flight from; it can tell us that the birds are distributed in the airspace at certain altitudes or at certain areas, such as all piling up along the coast versus being further inland; and it can tell us when are we getting big movements, like was March 1 the big spring migration date, or was it March 15."

Moon watching and flight call recording had given ornithologists of the 1950s a few tools to start observing bird migration patterns across time and space. But it took a young Louisiana biologist named Sidney Gauthreaux to recognize the potential of the new WSR-57 system for watching bird migration in a whole new way.

Growing up in New Orleans, Gauthreaux was interested in birds and bird migration from an early age. "In spring, every time the weather turned bad, migrants would rain down from the sky into woodlands," he told me by phone. In 1958, when Gauthreaux was in high school, he learned of plans to install one of the new WSR-57 radars on top of the New Orleans Federal Building. He wondered if this new technology might be able to detect birds in addition to incoming hurricanes, and his suspicions were confirmed when the unit was commissioned in November 1960.

By this time Gauthreaux was a sophomore at the University of New Orleans. "When they announced the installation of the new radar on television, I looked at the screen, and I swore to myself I could see birds moving on there, because I knew the behavior of the birds and I could see the echoes on the radar screen that they were filming for this news program," said Gauthreaux, now in his eighties, who speaks with a rich Cajun accent. "So I went to this new WSR-57 and talked to the meteorologists, that's how I got connected with radar."

According to Gauthreaux, the meteorologists he met were excited but skeptical about the idea that some of the unexplained echoes they were seeing on their new equipment were birds; knowledge of Lack's work during World War II was not universal among weather scientists working with the new radars. Gauthreaux finally convinced them by borrowing a page from George Lowery's book, taking them up on the roof of the building on a night when the moon was full. "I had them look at the disk of the moon to see the silhouettes of birds flying across it. And they were convinced from that point on that some of the anomalous propagation that they'd talked about before was in fact being produced by migrating birds."

Gauthreaux had met George Lowery and Bob Newman at a state ornithology conference in New Orleans when he was fifteen; both men, he recalls, were wearing coats and ties, the standard uniform for scientists in the 1950s even when they were doing field work. (He still has his notes from that day's bird-watching outing and reports that he saw an impressive eighty species.) Eventually, Gauthreaux would study under Lowery for his master's degree and PhD, working with WSR-57 radar for both. Unlike the local meteorologists in New Orleans, Lowery and Newman were aware of radar studies of bird movements in Europe, but they were initially skeptical of Gauthreaux's plans to use radar to study migration, and Gauthreaux had to start by documenting a correlation between moon-watching data and radar data to convince them that radar could be useful. His meteorologist friends "allowed me basically to use the radar like it was mine" whenever they weren't tracking dangerous weather, and he quickly picked up the basics of how to operate the system.

It took a WSR-57 radar twenty seconds to make one 360-degree sweep, so Gauthreaux was seeing blips moving on a screen every twenty seconds in real time, and the radar scans were also recorded on film as they came in. By comparing his radar data with moon-watching observations, Gauthreaux developed his own version of the migration traffic rate, estimating how many birds might be in a given flock showing up on radar and how many birds it would take to saturate the entire radar display.

His work also finally settled the debate between Lowery and George Williams about whether migrating birds really flew directly across the Gulf of Mexico. Even after Lowery published his moon-watching observations from the Yucatán in 1946, Williams had continued to argue that the songbirds Lowery and others observed over the Gulf were only there because they'd been blown out to sea by severe weather. "With the radar work," said Gauthreaux, "I could actually pick up the leading edge of these birds arriving on the northern Gulf Coast near dawn. So that sort of put the cap in the bottle and said, okay, trans-Gulf migration is real."

Weather radar even made it possible to directly measure the altitude of migrating birds for the first time, demonstrating that the birds arriving over the Gulf could be as high as sixteen thousand feet overhead (although sixty-five hundred feet or less is more typical).

At that time, there were no computer programs to filter birds out of radar images used for television weather reports. "Meteorologists in New Orleans, because of their interaction with me, started to actually say on television, this is not precipitation. This is migrating birds in the atmosphere," said Gauthreaux. "Ultimately the broadcast meteorologists of North America had a meeting at Disney World and asked me to give a presentation on my work, and I did, and a few of them went home and started acknowledging that in spring and fall a lot of the display that looked like light rain on the radar screen wasn't in fact light rain; it was a nice, clear, starry sky full of birds."

The weather radar network isn't the only type of radar that has proved useful to ornithologists. In the 1970s, researchers including Ron Larkin and Timothy and Janet Williams conducted studies using tracking radars—technology that might have been more familiar to Irven Buss and David Lack, because they're designed for tracking individual targets through space—at coastal sites and on seagoing research vessels in the Atlantic Ocean. Larkin and the Williamses found that in addition to the birds that cross the Gulf of Mexico, some birds were reaching South America by migrating over the open waters of the western North Atlantic, just as Ian Nisbet had controversially predicted blackpoll warblers must do based on banding data.

Later ornithologists, both in the United States and in Europe, would further refine the use of tracking radar for gathering data on the behavior, speed, and altitude of individual birds in flight, even mounting small radar systems on vans so that they could be easily moved from one location to another. Gauthreaux experimented with tracking radars himself, as recalled by a staff member at a lab on an island off the coast of Alabama in the 1980s: "I watched an RV with a marine radar unit attached to the top pull up in the parking lot [and] I thought, this guy really doesn't want to get lost. That's when I met Sid."

Today Gauthreaux is a professor emeritus at Clemson University. Although semiretired, he's still involved in radar ornithology research. "Some of my colleagues say, how can you sustain your interest in this when you're gonna be eighty years old next month? Well, if you're so excited and passionate about something, you just can't turn that off like a light switch. I think about it all the time, I'm excited about it," he says. "When I was an academician, I had other responsibilities, teaching courses, committee meetings, faculty meetings, and the like. Now that I'm retired, I can concentrate on the stuff I love the most. So I'm still at it."

The Radar Ornithology Renaissance

Radar ornithology hit a lull in the 1980s as bird researchers reached the limits of what they could do with existing weather radar networks. But when the last WSR-57 radar was decommissioned in South Carolina in December 1996, taking its place was something that would reinvigorate the field. The new network's official name was WSR-88D, but it came with a catchy nickname that reflected its futuristic capabilities: NEXRAD, for "next-generation radar."

There was one crucial upgrade: unlike its predecessor, NEXRAD is a Doppler radar system. The Doppler effect, named for the Austrian physicist who described it in the 1840s, describes how a wave's frequency changes depending on its motion relative to an observer. This is the phenomenon at work when you hear the pitch of a siren change as it passes you, sounding higher when it's approaching you than when it's moving away. As the source of the sound comes toward you, the sound waves between it and you are bunched together, and you perceive these higher-frequency waves as a higher-pitched noise. After it passes you, the sound waves get stretched out, with the opposite effect.

Although the Doppler effect is easiest to observe with sounds, the same principle applies to electromagnetic radiation, including radio waves. Unlike WSR-57 radars that only measured the angle and timing of returning radio energy, NEXRAD keeps track of the shape of the returning pulses themselves. If a returning pulse has been compressed, it means the object it bounced off is moving toward the radar antenna. If it's been stretched out, the target is moving away.

Now, in addition to seeing where a group of birds was at a fixed moment in time, ornithologists could easily track their speed and direction. Irven Buss and Sidney Gauthreaux could estimate the speed of passing flocks by tracking the progress of blips across a radar display, but NEXRAD collects this data automatically.

And in 2013, another, even more high-tech upgrade came to the U.S. weather system, called dual polarization. When a radio wave goes out, it oscillates in a single direction. (Think about holding on to the end of a piece of ribbon and shaking it side to side or up and down to create waves in different planes.) Without dual polarization, radar engineers must make a decision about which way to polarize the radiation, depending on what dimension they are more interested in "seeing": horizontally polarized radar shows the shape of objects in the horizontal plane, vertically polarized radar in the vertical plane. For meteorologists, horizontal polarization generally wins out, because a radar wave oscillating from side to side provides data about the changing shape of raindrops as they fall.

The 2013 upgrade, however, meant that for the first time weather radars operated in the vertical and horizontal planes simultaneously. That allowed ornithologists to determine the shapes of birds on radar in unprecedented detail—including which way they were pointing in the air. Although you could tell in what direction birds were moving with previous generations of weather radar, that direction was really a combination of where the bird was flying and where the wind was pushing that bird; imagine setting out to swim straight across a river but being pushed downstream by the current as you go. Being able to tell in what direction flying birds are actually oriented while this is happening has allowed ornithologists to answer new questions about how migrating birds drift on the wind.

Scientists have used weather radar to identify important "stopover" sites where migrating birds rest and refuel, see how they handle the need to fly over or around obstacles like the Great Lakes, and determine how strategies for successful migration differ between spring and fall. But with advances in radar technology have come advances in ornithologists' ambitions. Watching what masses of migrating birds were doing *now* is one thing. Meteorologists use weather radar to predict the future. What if ornithologists could do the same?

Bright Lights and Big Data

Imagine you're a ruby-throated hummingbird. It's May, and after a winter spent in the tropics, you're finally on your way home to the forests of eastern North America. Soon, instinct will compel you to build a nest, lay eggs, and hopefully pass your genes on to another generation. You've spent the night crossing the Gulf of Mexico, flying high above the dark water, wings beating nonstop for hours. The last of your energy reserves, built up in a feeding frenzy before you departed from the Yucatán, are nearly gone.

Ahead, dazzling lights glimmer on the horizon, and you adjust your course slightly to head toward them. Suddenly you're over land, but instead of a forest where you can rest and refuel, below you there is nothing but asphalt. The lights you were drawn to are the high-rise office buildings of downtown Houston, which loom all around.

Disoriented, you fly into one of them.

We don't really know why birds migrating at night are attracted to the artificial lights of cities. It may be related to how birds navigate, using the sun and stars as part of their suite of cues to help them orient in the right direction. But we do know that as many as a billion birds die in building collisions like this one every year in the United States alone.

The ideal solution would be to turn out city lights for the full length of the migration season and let the birds pass by in the darkness that evolution has prepared them for. But spring migration lasts for weeks, and shutting off the lights of downtown Houston and Dallas for weeks at a time is, sadly, not realistic. The birds don't arrive in a uniform stream over that entire time span, however. There are peaks and valleys in the number of birds arriving from the Gulf, as the tiny travelers adjust their schedule to take advantage of favorable weather conditions. What if we could predict in advance which nights would be quiet and which would see an avian traffic jam in the skies, and switch off the city lights exactly when it was most needed?

Enter BirdCast.

The BirdCast project began as an effort to protect migrating birds not from city lights but from pesticides. In the late 1990s, with funding from the EPA, a group of organizations and researchers including Sid Gauthreaux at Clemson University set out to develop a project that would use weather radar data to predict the intensity of bird migration in the "mid-Atlantic flyway," a region stretching along the East Coast from North Carolina to New England. The idea was that property managers could use information from BirdCast to avoid spraying potentially harmful pesticides when especially high numbers of migrating birds would be passing through.

It was a talk by Sid Gauthreaux at the 1998 meeting of the American Ornithologists' Union in St. Louis that drew a young ornithologist named Andrew Farnsworth to Clemson and set him on a path toward a career with radar and, specifically, BirdCast. A few years after finishing his bachelor's degree, Farnsworth was traveling and leading bird-watching tours while deciding what he wanted to do next.

After a summer on the coast of Texas, watching weather radar to help figure out when large numbers of migrants would be arriving on the coast, he headed north for the conference in St. Louis hoping to connect with potential advisors for graduate school. "Seeing Sid's talk—it was just the coolest thing ever," said Farnsworth, who recounted the full history of BirdCast to me over the course of multiple phone interviews. "He talked about how not only could you use radar to monitor the movements of birds, but you could look at their behavior in the air and assign their origins on the ground to certain habitats, and that was just a mind-blowing moment for me." After the presentation, Farnsworth approached Gauthreaux, expressing his interest in studying with him. Farnsworth arrived at Clemson just as the BirdCast project was starting, and he's still working on its latest iteration today.

That first version of BirdCast relied on a set of equations originally developed by Gauthreaux in the 1970s to predict the amount of bird migration each day based on weather conditions. Twice a day, Gauthreaux and Farnsworth or another graduate student gathered data from weather stations in the Atlantic flyway via the internet and input them into Gauthreaux's mathematical model to generate a forecast. To verify the forecast, they then downloaded NEXRAD radar data and created images showing the actual amount of migration in the region. Each morning and evening, the forecast, analysis, and images were posted to a public website that bird-watchers and anyone else interested in migration could access.

Doing all of this with turn-of-the-millennium technology was incredibly time consuming. "It required ingesting the data through a satellite dish on top of the biology building at Clemson, manually tending to the downloads, and then, the morning after, we would have to gather those data, assemble them in this weather visualization program, make this image, and then load it to the server," Farnsworth recalled. "If it wasn't there, there'd be this big hole on the project website. It was all pretty intense."

It meant Farnsworth always had to be near a good internet connection, at a time when internet access was far from ubiquitous. In the fall of 2000, the World Series between the New York Yankees and the New York Mets presented a challenge. According to Farnsworth, this was "basically the holy grail" for a New York baseball fan. Somebody in his family knew somebody who knew somebody who could get him last-row tickets for game five. So, Farnsworth completed the Thursday post before driving from Clemson to Atlanta to catch a flight to make it to LaGuardia in time for the game that night. After spending all night celebrating the Yankees' win at a local bar, "I went back to the airport for a 6:00 a.m. flight, got to Atlanta, got back in the car for a three-hour drive to Clemson, arrived literally just in time to post that Friday's forecast." But it was worth it: he invited a woman he barely knew at the time to go to the game with him, and today he's married to her.

Ultimately, BirdCast version 1.0 required an amount of labor that wasn't feasible to continue indefinitely, and the original BirdCast ceased its run in 2001. It took almost two more decades for computing power to catch up with the ambitions of ornithologists.

Farnsworth completed a PhD from Cornell University in 2007 and stayed on at Cornell first as a postdoctoral research associate and then in a permanent research position. The National Science Foundation (NSF) had funding available for projects that would apply advances in computer science to other fields, and Farnsworth became one of the co-primary investigators on a 2010 NSF grant proposal to resurrect BirdCast.

"By that point, big data was a concept that people understood," explained Farnsworth. "The phrase started to have meaning: we can take huge amounts of information and start to figure out how to process it. Computing power was advancing, and with that came increasing amounts of cloud-based data storage, and all of that was evolving really quickly. And all of those things were critically important to where BirdCast would go."

Radar data had become easier to access, too. Gone were the days of the NEXRAD archives being stored on tapes. "Even when I was a graduate student, I would have to request radar data for a certain time period, and then I'd wait a day or two, and then I'd get an email to download this big pile of data. My desk was just covered in hard drives, I had so many terabytes of data," Kyle Horton, a longtime Bird-Cast collaborator who's now an assistant professor at Colorado State University, told me. But in 2015, the U.S. National Oceanic and Atmospheric Administration partnered with Amazon Web Services to store the entire NEXRAD archive in the cloud and make it freely available for download. "The difference is that now the data are always on tap for you."

The next breakthrough came in 2018. Benjamin Van Doren, who had worked on the project as an undergraduate at Cornell, and Horton, then a Cornell postdoc, downloaded NEXRAD data for every single evening since the system was installed. These 150,000-plus individual radar scans spanned the entire continent over a period of twenty-three years. Van Doren and Horton then analyzed this massive data set to find out what weather factors—wind, air temperature, barometric pressure, and so on—predicted the appearance of big migration movements on radar. Essentially, they were following the path Sid Gauthreaux first laid out with the mathematical models he developed in the 1970s, but with a far more massive data set and greater computing power. (Although Horton never studied directly under Gauthreaux, in professor parlance he's Gauthreaux's "academic great-grandson": Gauthreaux was one of Horton's grad school mentor's mentor's mentor.)

---

* Ornithology research using weather radar in Europe has moved more slowly due to differences between countries in how weather radar data is collected, formatted, and stored, but research there is picking up as well. A 2019 study led by Cecilia Nilsson was the first to use weather radar to look at continent-wide migration patterns in Europe.

The main text of the resulting scientific paper, published in the journal *Science* in 2018, is only three pages long. But in it, Van Doren and Horton laid out a system for predicting mass movements of migrating birds on a continental scale. Their math explained almost 80 percent of the variation in migration intensity from one night to the next. "Having done analysis for a while now, when you're working a model in biology or ecology, you have certain expectations of what's good," said Horton. "When we got those results, we were like, oh, this is really good."

The most important factor, as it turned out, was air temperature, probably because of the relationship between air temperature and winds favorable to crossing the Gulf. Crucially, the model Van Doren and Horton created could predict migration intensity several days in advance, using only current weather conditions.

"That just seemed like a very powerful thing, both from the standpoint of getting bird-watchers excited, but also as a real tool to do conservation with," said Horton. "We could say, this is going to be one of the big nights in Texas or in Oklahoma or New York or wherever you are. We could predict it fairly accurately."

Based on their analysis of the NEXRAD archives, Van Doren and Horton concluded that during peak migration the number of birds on the move on a single night frequently exceed 200 million. Two hundred million is, for comparison, roughly the human population of Brazil. A nation of birds of all shapes and sizes, passing overhead as we sleep.

BirdCast might have gotten its start as an effort to protect birds from pesticides, but by this time BirdCast scientists were turning their attention to a different threat: city lights. Even though we're still not sure exactly why lights are so irresistible to migrating birds, radar is helping us unravel just how much of a threat those lights are. A 2018 study led by the researchers James McLaren and Jeff Buler used radar data to show that city lights actually affect migrating birds' habitat usage at a broad scale, with dense clouds of migrants descending on brightly lit cities across the country during migratory stopovers instead of settling in darker, potentially better habitat nearby. And in another big-data project that analyzed the entire NEXRAD archives,

Kyle Horton and his colleagues ranked the worst cities for exposing migrating birds to light at night (Chicago, Houston, and Dallas topped the list) and determined that half of a season's migrants typically pass through these crucial areas over just a handful of nights.

Campaigns in cities to encourage businesses and residents to turn out exterior lights during migration have been around since at least 1993. But Horton and his colleagues are working on refining these efforts, using BirdCast forecasts to identify the most critical nights, the nights when weather conditions mean that the numbers of migrants passing through will be exceptionally large. The hope is that city dwellers who may be reluctant to give up their lights for weeks at a time will still be receptive to these "lights-out alerts" and clear a darkened path for the millions of birds winging their way north on these special nights.

Residents of major cities along the Gulf Coast and elsewhere can sign up via the BirdCast website to receive alerts when a big migration night is imminent. In 2019, American National Insurance, based in the nearby coastal city of Galveston, pledged to turn out the lights at its headquarters building, which had proved fatal for migrating birds in the past. The U.S. Fish and Wildlife Service tweets out the #LightsOut hashtag to alert its followers in advance of big nights. Even the former first lady Laura Bush, who apparently became a bird lover at age ten when she earned her bird badge for Girl Scouts, has helped promote the campaign to turn off lights in Texas cities at the height of migration.

In the fall of 2019, a landmark study in the journal *Science* announced an alarming finding: North America's bird populations have declined by almost 30 percent since 1970, a loss of approximately three billion individual birds. NEXRAD data from 2007 to 2017 was part of the authors' analysis. The results from the radar data showed that the biomass of nocturnal migrants traveling along the continent's flyways each spring declined by 13 percent in that decade alone.

But there's reason to hope. Weather radar may be helping us watch birds' decline, but as we get better and better at pinpointing exactly where, when, and how we need to protect them during migration, it can help us save them, too. Irven Buss would surely approve.

Four
Follow That Beep

A Swainson's thrush looks a bit like a small brown version of its familiar cousin the American robin. Its gray-brown back contrasts with a pale, spotted chest and pale "spectacle" markings around its eyes. These thrushes are shy birds that forage for insects in the leaf litter on the forest floor, where they blend in with the dappled light and deep shadows. Birders know them by their fluting, upward-spiraling song, which fills the woods of Canada and the northern United States with ethereal music in summer. But they don't live there year-round; they spend the winters in Mexico and northern South America, then return north to breed.

On the morning of May 13, 1973, a Swainson's thrush pausing on its journey from its winter home to its summer home blundered into a mist net in east-central Illinois. The researchers who gently pulled it from the net went through all the usual rituals—weighing and measuring it, clasping a numbered metal band around its leg—but they added one unusual element: a tiny radio transmitter weighing just five-thousandths of an ounce. They carefully trimmed the feathers from a small patch on the bird's back, then used eyelash glue to cement the transmitter, mounted on a bit of cloth, in place against the bird's skin.

(Generations of ornithologists have learned exactly where to find the eyelash glue at their local cosmetics store. Designed to not irritate the delicate skin of the eyelids when attaching false eyelashes, it doesn't irritate birds' skin, either, and wears off after weeks or months.)

When the thrush was released, it probably shuffled its feathers a few times as it got used to its new accessory, then returned to resting and foraging in preparation for continuing its trek. At only around 3 percent of the bird's total body weight, the transmitter wouldn't have impeded the bird noticeably as it went about its daily routine. Then, around 8:40 that evening, after the sun had dipped far enough below the horizon that the evening light was beginning to dim, the thrush launched itself into the air, heading northwest.

It would have had no way of knowing that it was being followed. Bill Cochran—the same engineer who, a decade and a half earlier, had rigged up a tape recorder with a bicycle axle and six thousand feet of tape so that Richard Graber could record a full night of nocturnal flight calls—had been waiting nearby in a converted Chevy station wagon with a large antenna poking out of a hole in the roof. When the thrush set out into the evening sky, Cochran and a student named Charles Welling were following on the roads below. All they could see in the deepening night was the patch of highway illuminated by their headlights, but the sound of the wavering "beep . . . beep . . . beep" of the transmitter joined them to the thrush overhead as if by an invisible thread.

They would keep at it for seven madcap nights, following the thrush for more than 930 miles before losing the signal for good in rural southern Manitoba on the morning of May 20. Along the way, they would collect data on its altitude (which varied from 210 to 6,500 feet), air and ground speed (eighteen to twenty-seven and nine to fifty-two miles per hour, respectively, with the ground speed depending on the presence of headwinds or tailwinds), distance covered each night (65 to 233 miles), and, crucially, its heading. Because they were able to stick with the bird over such a long distance, Cochran and Welling were able to track how the precise direction the bird set out in each night changed as its position changed relative to magnetic north. The gradual changes they saw in its heading were consistent with the direction of magnetic north, providing some of the first real-world evidence that migrating songbirds use some sort of internal magnetic compass as one of their tools for navigation.

Today Bill Cochran is a legend among ornithologists for his pioneering work tracking radio-tagged birds on their migratory odysseys. But it wasn't birds that first drew him into the field of radio telemetry; it was the space race.

From Sputnik to Ducks

In October 1957, the Soviet Union launched the world's first artificial satellite into orbit. Essentially just a metal sphere that beeped, Sputnik 1 transmitted a radio signal for three weeks before its battery died. (It burned up in the atmosphere in January 1958.) That signal could be picked up by anyone with a good radio receiver and antenna, and scientists and amateur radio enthusiasts alike tracked its progress around and around Earth.

It caused a sensation around the world—including in Illinois, where the University of Illinois radio astronomer George Swenson started following the signals of Sputnik 1 and its successors to learn more about the properties of Earth's atmosphere. Around 1960, Swenson got permission to design a radio beacon of his own to be incorporated into a Discoverer satellite, the U.S. answer to the Sputnik program. In need of locals with experience in electrical engineering to work on the project, he recruited Bill Cochran (who still had not officially finished his engineering degree—he wouldn't complete the last class until 1964) to assist.

Cochran, as you may recall, had spent the late 1950s working at a television station in Illinois while studying engineering on the side and spending his nights helping Richard Graber perfect his system for recording nocturnal flight calls. By 1960, no longer satisfied with flight calls alone as a means of learning about migration, Graber had procured a small radar unit and gotten Cochran a part-time job with the Illinois Natural History Survey helping operate it. But along the way, Cochran had apparently demonstrated "exceptional facility with transistor circuits," which is what got him the job with Swenson. It was the transistor, invented in 1947, that ultimately made both the space race and wildlife telemetry possible.

The beating heart of a radio transmitter is the oscillator, usually a tiny quartz crystal. When voltage is applied to a crystal, it changes shape ever so slightly at the molecular level and then snaps back, over and over again. This produces a tiny electric signal at a specific frequency, but it needs to be amplified before being sent out into the world. Sort of like how a lever lets you turn a small motion into a bigger one, an amplifier in an electrical circuit turns a weak signal into a stronger one.

Before and during World War II, amplifying a signal required controlling the flow of electrons through a circuit using a series of vacuum-containing glass tubes. Vacuum tubes got the job done, but they were fragile, bulky, required a lot of power, and tended to blow out regularly; owners of early television sets had to be adept at replacing vacuum tubes to keep them working. In a transistor, the old-fashioned vacuum tube is replaced by a "semiconductor" material (originally germanium, and later silicon), allowing the flow of electrons to be adjusted up or down by tweaking the material's conductivity. Lightweight, efficient, and durable, transistors quickly made vacuum tubes obsolete. Today they're used in almost every kind of electric circuit. Several billion of them are transisting away inside the laptop I'm using to write this.

As transistors caught on in the 1950s, the U.S. Navy began to take a special interest in radio telemetry, experimenting with systems to collect and transmit real-time data on a jet pilot's vital signs and to study the effectiveness of cold-water suits for sailors. These efforts directly inspired some of the first uses of telemetry for wildlife research. In 1957, scientists in Antarctica used the system from the cold-water suit tests to monitor the temperature of a penguin egg during incubation, while a group of researchers in Maryland borrowed some ideas from the jet pilot project and surgically implanted transmitters in woodchucks.* Their device had a range of only about twenty-five yards, but it was the first attempt to use radio telemetry to track animals' movements. The Office of Naval Research even directly funded some of the first wildlife telemetry experiments; navy officials hoped that radio tracking "may help discover the bird's secret of migration, which disclosure might, in turn, lead to new concepts for the development of advanced miniaturized navigation and detection systems."

Cochran didn't know any of this at the time. Nor did he know that the Discoverer satellites he and Swenson were building radio beacons for were, in fact, the very first U.S. spy satellites; he and Swenson knew only that the satellites' main purpose was classified. Working with a minimal budget, a ten-pound weight limit, and almost no information about the rocket that would carry their creation, they built a device they dubbed Nora-Alice (a reference to a popular comic strip of the time) that launched in 1961. Cochran was continuing his side job with the Illinois Natural History Survey all the while, and eventually someone there suggested trying to use a radio transmitter to track a duck in flight.

"A mallard duck was sent over from the research station on the Illinois River," Swenson later wrote in a coda to his reminiscences about the satellite project. "At our Urbana satellite-monitoring station, a tiny transistor oscillator was strapped around the bird's breast by a metal band. . . . The duck was disoriented from a week's captivity, and sat calmly on the workbench while its signal was tuned in on the receiver. As it breathed quietly, the metal band periodically distorted

---

* Although harnesses, collars, and the like are also commonly used for tracking wildlife today, surgically implanting transmitters has its advantages, such as eliminating the chance that an external transmitter will impede an animal's movements.

and pulled the frequency, causing a varying beat note from the receiver."

Swenson and Cochran recorded those distortions and variations on a chart, and when the bird was released, they found they could track its respiration and wing beats by the changes in the signal; when the bird breathed faster or beat its wings more frequently, the distortions sped up. Without even meaning to, they'd gathered some of the very first data on the physiology of birds in flight.

An Achievement of Another Kind

Bill Cochran enjoys messing with telemarketers. So, when he received a call from a phone number he didn't recognize, he answered with a particularly facetious greeting.

"Animal shelter! We're closed!"

"Uh . . . this is Rebecca Heisman, calling for Bill Cochran?"

"Who?"

"*Is* this Bill Cochran?"

"Yes, who are you?"

Once we established that he was in fact the radio telemetry legend Bill Cochran, not the animal shelter janitor he was pretending to be, and I was the writer whom he'd invited via email to give him a call, not a telemarketer, he told me he was busy but that I could call him back at the same time the next day.

Cochran was nearly ninety when we first spoke in the spring of 2021. Almost five decades had passed since his 1973 thrush-chasing odyssey, but story after story from the trek came back to him as we talked. He and Welling slept in the truck during the day when the thrush landed to rest and refuel, unwilling to risk a motel in case the bird took off again unexpectedly. While Welling drove, Cochran controlled the antenna. The base of the column that supported it extended down into the backseat of their vehicle, and he could adjust the antenna by raising, lowering, and rotating it, resembling a submarine crewman operating a periscope.

At one point, Cochran recalled, he and Welling got sick with "some kind of flu" while in Minnesota and, unable to find a doctor willing to see two eccentric out-of-towners on zero notice, just "sweated it out" and continued on. At another point during their passage through Minnesota, Welling spent a night in jail. They were pulled over by a small-town cop (Cochran described it as a speed trap but was adamant that they weren't speeding, claiming the cop was just suspicious of the weird appearance of their tracking vehicle) but couldn't stop for long or they would lose the bird. Welling stayed with the cop to sort things out while Cochran went on, and after the bird set down for the day, Cochran doubled back to pick him up.

"The bird got a big tailwind when it left Minnesota," Cochran said. "We could barely keep up, we were driving over the speed limit on those empty roads—there aren't many people in North Dakota—but we got farther and farther behind it, and finally by the time we caught up with it, it had already flown into Canada."

Far from an official crossing point where they could legally enter Manitoba, they were forced to listen at the border as the signal faded into the distance.

The next day they found a border crossing (heaven knows what the border agents made of the giant antenna on top of the truck) and miraculously picked up the signal again, only to have their vehicle start to break down. "It overheated and it wouldn't run, so the next thing you know Charles is out there on the hood of the truck, pouring gasoline into the carburetor to keep it running," Cochran recalled. "And every time we could find any place where there was a ditch with rainwater, we improvised something to carry water out of the ditch and pour it into the radiator. We finally managed to limp into a town to get repairs made."

Cochran recruited a local pilot to take him up in a plane in one last attempt to relocate the radio-tagged bird and keep going, but to no avail. The chase was over. The data they had collected would be immortalized in a terse three-page scientific paper that doesn't hint at all the adventures behind the numbers.

That 1973 journey wasn't the first time Cochran and his colleagues had followed a radio-tagged bird cross-country, nor was it the last. After his first foray into wildlife telemetry at George Swenson's lab, Cochran quickly became sought after by wildlife biologists throughout the region. He first worked with the Illinois Natural History Survey biologist Rexford Lord, who was looking for a more accurate way to survey the local cottontail rabbit population. Although big engineering firms such as Honeywell had already tried to build radio tracking systems that could be used with wildlife, Cochran succeeded where others had failed by literally thinking outside the box: instead of putting the transmitter components into a metal box that had to be awkwardly strapped to an animal's back, he favored designs that were as small, simple, and compact as possible, dipping the assembly of components in plastic resin to seal them together and waterproof them. Today, as in Cochran's time, designing a radio transmitter to be worn by an animal requires making trade-offs among a long list of factors: a longer antenna will give you a stronger signal, and a bigger battery will give you a longer-lasting tag, but both add weight. Cochran was arguably the first engineer to master this balancing act.

The transmitters Cochran created for Lord cost eight dollars to build, weighed a third of an ounce, and had a range of up to two miles. Attaching them to animals via collars or harnesses, Cochran and Lord used them to track the movements of skunks and raccoons as well as rabbits. Cochran didn't initially realize the significance of what he'd achieved, but when Lord gave a presentation about their project at a 1961 mammalogy conference, he suddenly found himself inundated with job offers from biologists. Sharing his designs with anyone who asked instead of patenting them, he even let biologists stay in his spare room when they visited to learn telemetry techniques from him. When I asked him why he decided to go into a career in wildlife telemetry rather than sticking with satellites, he told me he was simply more interested in birds than in a job "with some engineering company making a big salary and designing weapons that'll kill people."

Cochran's first attempt at following birds on migration was unsuccessful. In October 1962, he and some colleagues fitted snow geese in South Dakota with transmitters, intending to track them south. The geese obstinately stayed put, however, and Cochran went home when the Cuban missile crisis took over the headlines and he became nervous about the idea of an impending nuclear war. But the idea stuck.

It was Cochran's old mentor Richard Graber, still intent on learning more about the songbirds migrating through Illinois, who first hit upon the idea of tracking thrushes. "We had listened to and recorded the calls of night migrants as they passed over Central Illinois," Graber wrote in a 1965 article for *Audubon Magazine*, "we had watched the migration on radar, [but] although we had learned something, the information always seemed so little. Some of it was decidedly confusing." Finally, he concluded that the only way to find answers to his questions was to "go right along with the birds" as they made their flights.

Graber picked thrushes for three reasons: they were nocturnal migrants (his area of interest), they were strong fliers (meaning they would be able, he hoped, to carry a transmitter), and he felt he knew enough about their behavior to predict which nights they would embark on big migratory flights. By 1965, Cochran, always building lighter and lighter transmitters, had gotten his creations down to eight thousandths of an ounce, about the weight of a dime. After weeks of experimentation, Graber came up with the method of gluing a transmitter to a thrush's back; when he observed captive birds with the glued-on transmitters, they preened and fed as normal, apparently unbothered.

On May 24, 1965, they captured a suitable candidate at their site outside Urbana—a gray-cheeked thrush, close cousin of the Swainson's thrush Cochran would follow years later. It took only ten minutes to band the bird, glue on the transmitter, and release it. That night, Graber and a pilot named Jim Taylor followed the bird for four hundred miles, two hundred of which were over the open water of Lake Michigan, in a plane they nicknamed the Porcupine because of all the antennas sticking out of it.

"Each of us, at times, must stand in awe of mankind, of what we have become, what we can do," Graber wrote in his piece for *Audubon Magazine*. "The space flights, the close-up lunar photographs, the walks in space—all somehow stagger our imagination. I was thinking about this as I flew south from Northern Wisconsin [the next morning], having just witnessed an achievement of another kind by another species."

The era of true migration tracking had begun.

Cochran would continue to build on this work in the ensuing decades, following falcons and eagles as well as thrushes by plane and by truck. (According to one article, his adventures over the years included "a near-plane crash, a hairy situation in which he was confused for a drug runner, and a thunderstorm so severe he radioed in his will.") As recently as 2015, researchers were using the same methods—the transmitter attached with fabric and eyelash glue, the truck with the giant antenna on top—to study the altitude changes thrushes make over the course of a night's flying. Bill Cochran's son James built the transmitters for that study, carrying on his father's work.

The 1973 trek from Illinois to Manitoba wasn't the farthest Cochran ever tracked a single bird; the following year he managed to follow a peregrine falcon all the way from Wisconsin to Mexico before having to give up due to "depletion of personal funds," as he put it in his published account of that work. But imagine Cochran and Welling's frustration when their epic weeklong chase of the thrush came to an end—hearing the "beep . . . beep . . . beep" fade away for the last time, the bird continuing into the Canadian night while they frantically tried to keep their failing truck running.

Fifty years later, researchers are still trying to follow thrushes. But now they can do it without leaving their offices.

Towers in the Andes

Swainson's thrushes spend the winter in the forests of Mexico and northern South America, lurking at the border between farmland and rain forest as they search for insects and berries, sometimes even following swarming army ants to snap up other insects that flee from their raids. In the Colombian Andes, many of them live on shade coffee plantations, where coffee plants are grown under a canopy of trees to more closely mimic the native ecosystem. On steep, misty mountainsides, thrushes mingle with Technicolor year-round residents like motmots and manakins in a mosaic of coffee plantations and scraps of second-growth forest. If the package of coffee in your pantry is labeled "shade-grown," your morning cup might have come from such a place.

Ana González grew up in these mountains, but she didn't grow up watching Swainson's thrushes. Like Bill Cochran meeting Richard Graber picking up dead birds at the base of a television tower, however, González altered the trajectory of her career forever when she discovered birds. As an undergraduate studying biology at a university in Colombia, the closest she got to birds was learning about the shapes of their bills via images on an overhead projector in a taxonomy class. Then a friend took her birding for the first time. Living in one of the most biodiverse regions of the world, they saw ninety species in a single morning. "I loved it," she remembered when she recounted for me via a video call how she first discovered ornithology. "I was in love."

González bought binoculars and started teaching herself about birds, eventually traveling to the Klamath Bird Observatory in Oregon to learn the art of bird banding and then to the Delta Marsh Bird Observatory in Manitoba to become their bander in charge. It was there that she connected with professors at the University of Saskatchewan, where she started her graduate studies in the ecology of Neotropical migratory birds in 2009.

For her PhD, González wanted to study how wintering migratory birds used the different habitats in the mountains of her childhood—

the shade coffee plantations as well as the second-growth forest inter-spersed with them. Swainson's thrushes, one of the most abundant species at the sites she was interested in, were a natural choice to work with.

When she began her PhD field work in 2015, González planned to use two different methods for following the movements of her birds. To keep track of them over the winter and record when they left the area in spring, she would use radio telemetry. And to follow their journeys north and analyze whether their migratory behavior was related to what habitat type they'd wintered in, she hoped to use a different tracking device called a light-level geolocator (more on those in another chapter). Ultimately, she didn't get enough data from the geolocators to carry out the analysis she'd hoped to do, but the ra-dio telemetry part of her research plan, despite a rough start, came through in a big way.

Officially launched in 2014 by the Canadian bird conservation or-ganization Bird Studies Canada (now called Birds Canada), the Mo-tus Wildlife Tracking System is a network of towers, each mounted with one or more antennas, that spans large swaths of North America and has a few outposts on other continents as well. The receiver tow-ers can pick up the signals of any animal (the system has been used for bats and large insects as well as birds) with a Motus transmitter that passes within range. Any qualified researcher can buy and put up towers in the region they're interested in, but if one of their tagged animals flies out of their study area and passes by a tower installed as part of a different research project somewhere else, they'll be notified about that as well. The name Motus comes from a Latin word for movement or motion.

González had recently received a fellowship from Bird Studies Canada and planned to use Motus radio telemetry for keeping tabs on her wintering thrushes. Her first challenge, however, came when her Motus receivers arrived in Colombia as disassembled collections of electronic components in boxes. She and her colleagues, despite lacking Bill Cochran's expertise in electronic engineering, had to put them together and get them to work. Getting them in place was challenge number two: one of the two sites where they wanted to erect a Motus tower was on a hard-to-access mountaintop, requiring an hour-long trek with a solar panel and two fifty-pound batteries on the back of a horse.

It took them about three weeks to get everything set up. Once they got going, however, González and her colleagues tagged a total of 284 birds across their two sites between 2015 and 2018, catching birds January through March and attaching the tiny transmitters to the birds using harnesses made of elastic thread, designed to wear out and fall off after about eight months. González wanted to find out whether birds captured in forest habitat, moister and richer with food resources, left for their spring migration earlier than birds that wintered in shade-grown coffee habitat, able to bulk up for the journey faster. She would be able to tell when each bird departed based on when her Motus receivers stopped detecting it in the area. She also originally hoped to follow the birds' winter movements in more detail, chasing after them with handheld receivers to figure out exactly how big each bird's home range was, but the steep Andean terrain made that part of her plan impossible.

In April 2015, the first of González's tagged birds headed north. Although she couldn't literally hear the beeps of their transmitters fading away, she felt something akin to what Bill Cochran experienced on that lonely Manitoba roadside forty years earlier as the thrushes she had come to know stopped being detected at her receivers. Their long journey home had begun.

The Ultimate Altruistic Adventure

The website of the Motus Wildlife Tracking System, the system González used to track her thrushes' departures from Colombia, describes it as "an international collaborative network of researchers that use automated radio telemetry to simultaneously track hundreds of individuals of numerous species of birds, bats, and insects."

Ask Stu Mackenzie what Motus is, however, and his first response is "the ultimate altruistic adventure in wildlife tracking."

In 2007, when I was a nineteen-year-old volunteer at the Long Point Bird Observatory, Stu was the guy who ran the show at the field station there. Today his official job title is director of migration ecology for the organization Birds Canada, but a big part of his job involves overseeing the Motus network. As he puts it, he's Motus's "grand congealer," responsible for keeping the sprawling collaboration organized and, as he wearily emphasized, making sure everyone gets paid. When I reached him for a video call, instead of sitting at a desk like most people I talked to for this book, he was walking around his yard in Port Rowan, Ontario, in sunglasses and keeping an eye on his kids, who were home from school due to the COVID-19 pandemic.

The idea to automate radio telemetry goes back to (surprise, surprise) Bill Cochran. After Cochran's success in Illinois with Rexford Lord, a University of Minnesota ornithologist named Dwain Warner offered him a job at the university's Cedar Creek Natural History Area getting a radio telemetry system up and running there. Like so many other scientists working at the time, Warner had been inspired by Sputnik to try to develop radio tracking technology for birds and other wildlife. In 1963, Warner and Cochran built a system of two tower-mounted receivers that could cover the entire forty-five-hundred-acre nature area at once, capable of following the movements of more than fifty animals simultaneously.

That system, of course, could track animals only as they went about their daily routines; it was useless for studying migration, immediately losing track of any creature that left the confines of the nature reserve. But although Cochran eventually returned to Illinois to track thrushes with Richard Graber, the idea of an automated radio telemetry network lived on.

In 2008, an array of three towers, each with multiple antennas mounted twenty-five to forty-five feet above the ground, was erected at Long Point. Long Point, as its name suggests, is a long, narrow peninsula of land that extends into Lake Erie from the shore of Ontario, attracting large numbers of migrating birds as they cross the open water of the lake. Coincidentally, the project was being set up the summer I was there, although I was completely unaware of it at the time.

First used to study the movements of Swainson's thrushes and their cousins the hermit thrushes as they passed through the area, then a wider range of species the following year, the Long Point array showed that birds on a "stopover" (a multiday rest stop between long-distance migratory flights) move around a lot more than researchers thought, sometimes traveling almost twenty miles overnight within their stopover area before finally moving on for good. Another array operated on a peninsula on Alabama's Gulf Coast and on Mexico's Yucatán Peninsula from 2009 to 2013, letting researchers track the Swainson's thrushes and other birds as they began and completed their migration across the Gulf of Mexico. (Bill Cochran built some of the transmitters for that study.)

But what if these researchers working at different locations along Swainson's thrushes' migratory path used the same transmitters and receivers? What if they could detect each other's birds, picking up a bird tagged at Long Point when it arrived in Alabama a few days later?

Motus got its start in 2012 at Acadia University, a small university in Canada's Atlantic coast province of Nova Scotia. Previously, the biggest barrier to setting up large networks of automated receivers was the cost of the equipment; each receiver could cost as much as ten thousand dollars. But Phil Taylor, the Acadia University ornithologist who had also overseen the development of the three-tower array at Long Point, developed a low-cost, open-source receiver he dubbed Sensorgnome, which could be built from off-the-shelf components for a fraction of the cost of the proprietary technology Taylor and his colleagues had been using previously.

In 2014, a major grant from the Canadian Foundation for Innovation gave the fledgling Motus program the resources it needed to scale up in a big way. It was at that point that Bird Studies Canada stepped in to oversee the growing program in collaboration with Acadia University and a range of other institutions. By the end of 2016, more than

120 separate research projects were using transmitters on the Motus network, which had expanded to 325 receiver stations in eleven countries, including a handful in Europe.

Motus's open-source nature is what makes it unique: tags deployed by any one research group can be picked up by receivers put up by any other researcher in any part of the world if a bird happens to fly by one. As Stu Mackenzie put it, "Everything about it is collaborative. Whether you like it or not, you're part of the family. You're contributing to other people's projects, and other people are contributing to your projects."

The smallest "nanotags" used by the Motus network weigh just seven-thousandths of an ounce, small enough to be used on almost any bird as well as many large insects, but the tiny batteries required to make such light transmitters last only a few weeks. Larger tags incorporating longer-lived batteries can stay active for many months, and new solar-powered tags can keep transmitting almost indefinitely.

But for the data collected to be useful, each of the tens of thousands of transmitters sharing the Motus network of receivers needs to be individually identifiable. You need to know whether that tag that was just picked up by a Motus tower in eastern Canada was one of the white-crowned sparrows in your study, or a green darner dragonfly being followed by a group of entomologists at the university two towns over. To make it work, each Motus transmitter sends out its own unique sequence of beeps—a sort of signature, beamed out by precisely one tagged animal.

Even as they passed out of range of the receivers Ana González had erected in Colombia, each of her tagged thrushes was continuing to stream out its own signature sequence as it winged its way north.

González knew there were other Motus towers put up by other researchers in Central and North America. Still, she'd accepted that she would likely never hear from any of her thrushes again once they left her study area. When she started her project, there were fewer than three hundred receivers in the network. "What are the chances that you would tag a Swainson's thrush in Colombia," she asked, "and this tiny bird is going to fly right next to these antennas in North America?"

Then, on May 25, 2015, about a month after her thrushes had left Colombia, González received an email from Stu Mackenzie. A thrush she had tagged in March had been picked up by a tower on Chaplin Lake in Saskatchewan—3,700 miles from where it had started out in Colombia. The Chaplin Lake Motus tower had been erected earlier that year by researchers studying shorebirds passing through the area. Incredibly, it was less than 125 miles from the University of Saskatchewan campus, where González was a PhD student. González and her colleagues in Colombia were astounded. And that was only the beginning.

Over the four years of her project, forty-four of the more than two hundred thrushes González tagged would be detected after leaving Colombia—some near the end of their journey in Canada, but others along the way as they passed by Motus towers in Central America and along the Gulf of Mexico. "Every time that we learned about a detection, it was a celebration," she told me—not just for the scientists, but also for the farmers on whose land they had been working, the bus driver who helped them get to their research sites, the housekeeper at the field station where they were staying. After following the lives of these individual birds over the winter, everyone was invested in their journeys north.

Ultimately, González was able to use the Motus detections to answer the question she'd originally hoped to tackle with light-level geolocators: Was their migratory behavior related to which habitat type, shade-grown coffee plantations or second-growth forest, they had spent the winter in? Based on past studies done with warblers wintering in the Caribbean, she expected that birds that spent the winter in shade-grown coffee plantations would leave Colombia later than forest-wintering birds. Coffee plantations offered less food for thrushes than forests, and González reasoned that the birds there would need extra time to build up enough fat reserves for the journey. As a result, González predicted, coffee plantation birds would arrive later on their breeding grounds and be at a disadvantage when it came to claiming the best nesting territories and raising the healthiest babies.

Instead, she found the opposite: not only did forest birds leave later, but they were able to "catch up" with coffee-dwelling birds by traveling faster, arriving in Canada at roughly the same time. González thinks that maybe the coffee-dwelling birds leave earlier because they know that after a winter spent in less-than-ideal habitat, they'll need more time to complete the grueling journey north.

The results only stoked González's desire to learn more. "I want to go back in April and the beginning of May and see how much fat these birds are gaining," she said, to untangle the relationship between habitat, fat reserves, and migration timing once and for all. Migration research is like a fractal. The closer you look, the more there is to see.

Back to Where It All Began

Once, the floodplain of the Illinois River, a short distance from Bill Cochran's old stomping grounds, would have been lined with natural wetlands fed by spring floods. Now penned in by levees, today these wetlands are filled with exotic mute swans and invasive phragmites reeds, but they still provide crucial habitat for migrating birds.

On a humid September day in 2021, the sunlight tinged yellow by smoke from distant California wildfires, I joined my friend Auriel Fournier as she checked her rail traps in the Nature Conservancy's Emiquon preserve. (Emiquon would have been cornfields in Cochran's heyday; the Nature Conservancy purchased the land in 2000 and has since been restoring the natural floodplain as best it can.) Fournier is the director of the Illinois Natural History Survey's Forbes Biological Station, the "research station on the Illinois River" from which that first radio-tagged duck in the 1960s came. At Forbes, Fournier also oversees research projects on the area's ducks and geese and even reptiles and amphibians, but rails—small, secretive wetland birds distantly related to cranes—are her real passion. She's been studying them since she was a PhD student and even goes by the handle @RallidaeRule on Twitter, an homage to the rail family's Latin name.

For the most part, rails pass through this part of Illinois only during migration; they don't breed here. But they make much longer stopovers during migration than songbirds, sometimes hanging out in the area for weeks in spring and fall before moving on. Fournier wants to know how they're using the local landscape during that time. Does each individual stay within a small area, or are they moving between different habitat types across a mosaic of private and public lands? The best way to manage habitat to benefit them depends on the answers, which Motus tags can provide. And of course, as a bonus, using Motus tags means Fournier sometimes finds out what happens to "her" rails after they leave Illinois. In the spring of 2021, her first season of tagging, birds she captured went on to be detected near Madison, Wisconsin, and Minnesota's Twin Cities.

As we moved toward each trap, pushing the tall wetland vegetation aside as we approached the wire cage and the speaker blaring recorded rail calls to lure birds in, we looked eagerly to see if this one might have an occupant. But each one was empty, as they had been the previous day. I had come to Illinois to see her put a Motus tag on a rail, but for that to happen, we'd have to catch one, and this was my last day in the field with her.

Fournier had one more rail-catching method to try—the one she'd used for her PhD field work in Arkansas. And she was determined not to send me home without having seen a rail get fitted with a transmitter.

At dusk, we met her crew of assistants back at the edge of the wetland. Fournier distributed spotlights and hand nets she'd bought earlier that day at Walmart while she checked how much fuel was in the tank of the ATV that had been left in the dirt parking area. The plan was that she would slowly drive the ATV into the vegetation while the rest of us fanned out on either side on foot. Any rails in the vicinity would flush at the sound of the approaching engine, flying up into the air and revealing their position before landing a short distance away. Each time a bird flushed, those of us on foot would note where it landed and run after it, capturing it safely in a hand net.

I'd tagged along on plenty of ornithology field work before, but never anything quite like this. We started into the wetland as darkness fell, pushing through chest-high grasses. My headlamp kept slipping down my forehead and distracting me as I struggled to concentrate on watching for flushing birds. Mosquitoes appeared out of the gloom in droves and bit me through my thin T-shirt.

"There! There!" Suddenly a small bird burst from the plants, highlighted in the glow of the spotlights, five people clutching nets sprinting full tilt after it. In moments it was safely immobilized and Fournier climbed down from the ATV to retrieve it.

The bird was a sora, a black and gray species of rail with a yellow bill and black face that can be found in marshes throughout much of North America. Fournier and one of her technicians spread out their tools on the back of the ATV. As I watched, they banded the bird, did some checks to make sure the Motus transmitter was functioning properly, and then attached it via a tiny harness that looped around the sora's legs and held the transmitter snug against the center of its back. After a quick pause for photos, the bird, now with an antenna extending past its tail, was released, and we all beat a quick path back to our vehicles to escape the mosquitoes.

See a bird get tagged with a Motus transmitter—check. (That hapless rail would later get picked up by receivers around Lake Erie, well to the east of where it should have been heading, which Fournier described as "interesting and unexpected, but pretty cool.") But I wasn't quite done with Illinois yet. I had one more stop to make.

The next morning, I picked up a rental car and made the ninety-minute drive to Champaign. Down a quiet side street not far from the University of Illinois campus, I pulled up in front of a house with an overgrown front yard and a suspiciously familiar-looking old Chevy station wagon in the driveway. My first knock resulted only in the sound of yapping dogs from inside, but when I tried again, a home health aide opened the door.

"I'm here to see Bill Cochran. He's expecting me," I said. Darn it. I'd hoped he'd answer the door himself so I could say I was looking for the animal shelter.

In a few moments I found myself sitting on the back porch with the man himself. A tall, almost gaunt old man, he said he wished I'd been able to visit a few years earlier when he was in better shape. His backyard was dominated by a pair of enormous antenna towers, taller than the house itself. (At first I thought they might be for ham radio, a hobby of Cochran's, but he set me straight: they were definitely for tracking birds.)

Periodically interrupting himself to point out sparrows flitting to and from his birdbath, Cochran regaled me with more stories about his adventures tracking birds, ones I hadn't heard yet despite two previous conversations by phone, like the time in the 1970s when he was mistaken for a smuggler. After tracking a falcon toward Cuba in a Cessna, he and his pilot turned back, unable to enter Cuban airspace, and landed at a small airstrip in the Florida Keys. "So we came in and we landed on the key, and we were surrounded by police, state troopers, probably national, I don't know, it must have been a hundred people out there with guns." I can't verify this story. But that's how Cochran remembers it.

Inside (masked and keeping our distance, this being during the COVID pandemic), he opened file cabinet drawers to pull out photocopies of obscure journal articles describing his exploits and wrestled with the search function on his desktop computer to show me videos shot inside the station wagon during a night of tracking. He was missing a daily check-in with his ham radio club to talk to me; at one point the phone rang, and although the answering machine picked it up before he could locate his cordless phone in his cluttered house, it turned out to have been one of his ham radio buddies calling to check up on him. So, he briefly switched on the radio and broke into their conversation to inform them that he wasn't joining them because he was too busy "talking to a beautiful woman" about his birds. ("You could probably talk for two weeks about your birds," I heard someone respond.)

In one of the stories Cochran told me, he and a buddy were resting in the shade under their plane after having made an excursion off the Florida coast to check the progress of a migrating peregrine falcon. "All of a sudden here comes this jet—zoom. It lands and out comes a party to greet them, you know, and all these guys with their briefcases get out. They were probably building a supermall or something like that. I remember I said, damn it, we're in the wrong business. But [my friend] said, no we aren't. We're having fun!" He laughed. "It was always an adventure. And it was always fun."

The Future of Motus

As of the summer of 2022, there were 502 individual research projects in the Motus database. Together, those projects had erected 1,381 individual receiver stations and deployed tags on 32,114 individual animals representing 284 species, including numerous birds as well as more than 30 bats, 4 bumblebees, 3 dragonflies, 1 butterfly, the northern quoll (a small polka-dotted marsupial tagged by a project in Australia), and, oddly, humans. (Presumably the 5 humans listed in the Motus database represent researchers goofing around and testing out their equipment.)

And the network is growing all the time. The greatest densities of Motus towers are in the Western Hemisphere, especially North America, where there is even a pair of towers on remote islands at the north end of Hudson Bay, at the edge of the Canadian Arctic. But the technology is catching on in Europe and Australia as well, and a pair of towers went up in Taiwan in 2019 to gather data on migratory shorebirds off the island's west coast. The bird conservation nonprofit BirdsCaribbean launched a new Caribbean Motus Collaboration in the spring of 2021, with a plan to install and maintain Motus receivers in strategic spots throughout the islands. Another researcher I spoke with was working on erecting Motus towers on the northwest coast of Mexico, hauling heavy equipment up tall poles to learn more about the under-studied shorebirds that winter in the region.

Stu Mackenzie would love to see more Motus towers go up in Europe, and he also hopes to eventually see the network expand into Africa. Ultimately, his dream is to see Motus become as mainstream as bird banding. "The potential is just vast," he said.

Someday you may even be able to install a Motus receiver in your backyard for no more than the cost of a tablet computer. In 2021, a company called Cellular Tracking Technologies ran a successful Kickstarter campaign for a device they dubbed Terra, promising that anyone who pledged at least $165 would receive a small disk-shaped gadget to place in their backyard that would both detect Motus-tagged birds in the area and monitor and identify nocturnal flight calls. As of the summer of 2022, the Terra Project website didn't list an estimate of when the devices would be ready to ship, but I'd like to imagine owning one someday.

In eastern Washington state, I live on the margin of the breeding range of the Swainson's thrush. But every now and then, walking near the lake east of town in summer, I hear that ethereal upward-spiraling song drifting through the cottonwoods. In what tropical forest did those birds spend the winter? What route, through what dangers, did they take to get here? I wonder if I've ever heard a radio-tagged bird singing and didn't know it. The chances are slim, but as Ana González learned, sometimes unlikely things happen. Maybe someday I'll be able to buy a receiver for my backyard and find out for sure.

Five
Higher, Further, Faster

Bird migration doesn't usually make headlines. But in September 2007, when the U.S. Geological Survey put out a press release with the title "Bird Completes Epic Flight Across the Pacific," a shorebird named E7 became an instant celebrity. Suddenly Brian Williams was talking about migration on the NBC evening news.

E7 was a bar-tailed godwit, a bird about the size of an American football, with long legs and a long, slightly upturned bill. Most of the world's bar-tailed godwits breed in northern Europe and Asia, but a few turn up to nest in Alaska every summer instead. Thanks to banding records, scientists knew even before E7 that the Alaska-breeding birds were spending the Northern Hemisphere's winter in New Zealand and Australia—an incredible migration, no matter the route. Recoveries of banded birds suggested that when making the trek north to nest, these godwits made a long hop to the east coast of Asia, followed the Asian coast north, and then crossed from Siberia to Alaska.

But records of bar-tailed godwits from Alaska were curiously absent on the coast of Asia in the Northern Hemisphere's autumn, when they should be making their way back south. The only explanation seemed to be that in the fall these birds must journey from Alaska to

New Zealand in a single long, unbroken flight across the full breadth of the Pacific. If true, it would be an almost-impossible-to-believe feat of migration.

Hoping to find out for certain what was happening, a group of scientists from the USGS and other organizations decided to follow the godwits' migration with transmitters. In 2006, they captured four females during their short breeding season in Alaska. Concerned about the possibility that attaching the bulky transmitters with external harnesses could affect the birds' aerodynamics during their long overwater flights, they surgically implanted the transmitters in the birds' body cavities instead, setting up mobile veterinary surgery units in tents on the tundra. One bird's transmitter failed before it left Alaska; one bird made it as far as an island in the South Pacific called Ouvéa but died there. The two remaining birds were on course for New Zealand but still more than nine hundred miles short of their presumed destination when their transmitters' batteries died. The researchers had hoped that the batteries would last longer, but in those early days of using implants to track shorebirds, they weren't sure what to expect.

Undeterred, the scientists made their own trip across the Pacific to New Zealand to tag a fresh batch of godwits at the other end of the world. They could at least gather new data on godwits' trip north through Asia before trying again to document the southward leg of their migration the following year.

E7 was one of the birds captured in New Zealand. As expected, when she left for her spring migration, she first crossed to Asia and traveled up the coast before returning to Alaska. But remarkably, her transmitter continued functioning throughout her short five-week breeding period in Alaska and into the beginning of her return trip south.

Lee Tibbitts, a USGS ornithologist who was part of the project, was kind enough to get on Zoom to reminisce about it for me (though she noted it was the third time that week she'd told the famous E7 story). "It was very exciting for us, because this was the first time we'd tracked shorebirds," she said. To save power, the transmitters were programmed to turn on for just six hours at a time after thirty-six hours off, and she and her colleagues kept close track of when they might expect a new location fix. "We'd gather around my computer and say, okay, she should report this hour!"

For nine days and nights, they watched E7 move south out of Alaska and across the open waters of the Pacific Ocean. When she touched down in New Zealand, she'd covered more than seven thousand miles, the longest nonstop migratory flight ever recorded.

Some seabirds, such as albatrosses, circumnavigate the world almost routinely; their wings, perfectly adapted for gliding on even the slightest wind, allow them to do this while expending very little energy. But a godwit is built for flapping. According to Tibbitts, part of what makes their extreme migratory feat possible is strong weather systems that move east from Japan each fall. "When they pass Alaska, there's a really nice tailwind going south," she said. The birds "jump on those tailwinds and just shoot out of Alaska. They're going a hundred kilometers [sixty miles] an hour."

Today E7's journey has been recounted in everything from documentaries to picture books, her achievement lauded on both sides of the Pacific. When Tibbitts and her colleagues traveled to China for a shorebird festival a few years later, she recalled trying to converse with a taxi driver ferrying them from their hotel to a restaurant. "He couldn't speak English very well, and I couldn't speak Mandarin very well, but we were trying to talk. He asked where we were from, and I said Alaska, we're here for the shorebirds. And he said, you mean like E7?"

The device that E7 carried was, of course, more than just a radio transmitter. No antenna-carrying pickup truck could follow her across the ocean; no network of tower-mounted receivers could cover the entire vastness of the Pacific. To track the world's most extreme bird migrations—the longest distances, the highest altitudes, the most endangered species—you need a different perspective, one that can take in the entire globe.

You have to get into orbit.

Receivers in Space

By now, you should not be surprised to learn that it was Bill Cochran who first came up with the idea of tracking animals via satellite.

Cochran and Dwain Warner, the Minnesota ornithologist who hired him to build the two-tower automated radio telemetry system at Cedar Creek, thought up the idea over beers at a conference in 1962. Reminiscing about his work building radio beacons for the Discoverer satellites, Cochran commented that his designs were so good he could put a radio-tagged rabbit on a satellite and be able to hear its beeps from Earth—or, he suddenly realized, "we can put the receiver in orbit and hear a rabbit from up there!" Cochran and Warner pursued the idea as far as making an official pitch to NASA, but their proposal was rejected. Cochran's ambitions outstripped the technology available at the time, and NASA might have been a bit preoccupied with other things: 1962 was the year it launched the first American into orbit, and the Apollo program quickly followed.

In 1970, a cow elk near Yellowstone National Park became the first animal to ever be tracked via satellite telemetry, when the twin brothers and wildlife biology legends Frank and John Craighead outfitted it with a radio collar that communicated with a weather satellite called the Nimbus 3. But elk are huge. Building a transmitter powerful enough to talk to orbiting satellites but small enough to put on a bird would be a far greater challenge.

The ornithology career of Bill Seegar, the man who would ultimately be responsible for what would be dubbed "the Bird-Borne Satellite Transmitter and Location Program," began with a PhD dissertation on the life cycle of a heartworm that parasitizes swans, perhaps an unlikely origin story for someone who went on to work with satellites. But after completing a research fellowship at a wildlife reserve in the U.K., Seegar took a job with the Department of Defense (DOD) in 1981. The DOD is one of the biggest landowners in the United States, managing roughly ten million acres. The military needs large tracts of open land to practice maneuvers at real-world scales, but it's also responsible for managing wildlife that lives on that land. Seegar's new bosses needed a way to monitor birds and other wildlife that didn't require having a bunch of biologists running around on bombing ranges in person.

Seegar was familiar with the Craighead brothers' work, but it was a chance dinner invitation in 1981 that ultimately made him a key figure in the history of wildlife telemetry. A friend of Seegar's invited him to come with him to dinner at his fiancée's house. At dinner, Seegar started talking about what he did for a living and his goal of developing new ways to track and monitor birds. The friend's fiancée's father just happened to be a lawyer for Johns Hopkins University's Applied Physics Lab (APL) and know the exact engineers who had just designed a new satellite-based navigational system for the U.S. Navy, and he offered to make some introductions.

"Now, I've got a degree in tropical medicine, I was not an electrical engineer," Seegar told me. "[But] they set up an opportunity for me to come down and present something to a roomful of engineers. And so I put together a set of nominal requirements: we want to be able to track something on a global scale, we want to be able to track something that's small, bird-sized, and we want to be able to track it for at least eight or nine months." His pitch was thorough, but there was one aspect he hadn't considered. "It wasn't long before somebody in the back of the room raised his hand. I said yes, and he said, Dr. Seegar, do you have eighty million dollars?"

Seegar did not have eighty million dollars. But he was able to scrape together enough funding from the DOD and other sources to at least do a feasibility study, and the engineers took it from there.

One of the newly formed team's first tasks was figuring out if any receivers already in orbit would work for what they had in mind. When Seegar and his new colleagues began to investigate whether any existing satellite systems might be suitable for tracking birds, their research immediately led them to one called Argos. Launched in 1978 as a collaboration between NASA, the U.S. National Oceanic and Atmospheric Administration, and France's National Center for Space

Studies, the Argos system works by making clever use of the Doppler shift. As a satellite approaches the location of a transmitter, the radio waves it receives get scrunched together and shift toward a higher frequency. Once the satellite passes over and begins to move away, the radio waves get stretched out and shift toward a lower frequency. By listening to the shifts in frequencies from an individual transmitter and combining that data with the satellite's speed and position, one can calculate the approximate position of the transmitter on Earth's surface.

Location fixes from Argos are accurate to within 250 meters, or about 820 feet. The Argos system was originally intended for keeping track of weather buoys and other oceanographic and meteorological devices, but to the surprise of its developers, wildlife scientists had co-opted the Argos system almost as soon as it was launched, first experimenting with tracking the movements of polar bears off the coast of Alaska. Surely it could be made to work for birds as well.

But they still had to build the transmitters. The APL team decided early on to design their transmitter to be powered by solar cells, since there was no battery available that was both lightweight and long lasting enough for their purposes, and experimented with putting solar cells on the roof of their building to see how much power they could generate. Getting accurate locations from the Argos system also required extremely stable transmitter frequencies, so they had to develop a way to compensate for temperature fluctuations that could throw off the frequency of a crystal oscillator. The device they eventually came up with weighed about six ounces, roughly the size and weight of a hockey puck. In May 1984, after trying it out on captive golden eagles, they decided it was time for the first field test.

They didn't go far at first—just to an island in Chesapeake Bay, about thirty miles from the lab. There, they captured a mute swan, an exotic species introduced from Europe. "You could hear that bird going *pat pat pat pat pat* as it was taking off from the water, and when those little feet stopped, and you didn't hear a thing, that was the first satellite transmitter that ever went into the air," said Seegar. "It was a small move for mankind, but a great move for avian conservation." For months, they were able to get location fixes on the bird every day as it raised its young on a pond on the island.

Next Seegar and his colleagues tested their devices on a bald eagle captured on the Maryland coast, and on trumpeter swans captured in Alaska when they were rendered flightless during their annual molt. (A report described the team's comical swan-catching technique in the typical dry language of scientific writing: "In order to capture a swan, a small pontoon plane with two or three people aboard lands on the pond. Holding a long net, one person stands on a pontoon; the pilot controls the plane and chases the bird around the pond until it is caught.") They were able to track two swans into Canada that fall before both birds apparently died, and the eagle's transmitter functioned for nine months as the bird moved up and down the Atlantic coast. That was about as far as Seegar's initial funding got them, but they had proved that tracking birds long distances from space was possible.

A trumpeter swan like the ones the APL team chased around ponds in Alaska can weigh as much as twenty-six pounds, about the same as the average two-year-old child. A bar-tailed godwit like E7 can weigh over a pound when bulking up for migration. But one of the world's most endangered—and cutest—shorebirds tops out at just under an ounce and a half, roughly the weight of fourteen cents in pennies. Luckily, one of Seegar's colleagues has gone on to build satellite transmitters a fraction of the size of those original 1980s prototypes, and they're becoming a key tool in the race to save this species.

Saving the Spoonie

The spoon-billed sandpiper is a small, plump bird, reddish in summer and gray-brown in winter, that lives along the ever-shifting margin where land and ocean meet. Migrating between Arctic Russia and southern Asia, it doesn't look that different from related shorebird species that you might see on the beaches of North America or Europe, with one striking exception: its beak. The feature that gives it its name is the unique spatula-shaped tip of its bill, and although ornithologists aren't completely sure of its purpose, it presumably helps in some way with foraging.

Spoon-billed sandpipers (or spoonies) were likely never abundant to begin with, but in the early years of the twenty-first century a group of Russian ornithologists sounded the alarm that these birds, which had not been thoroughly surveyed for three decades, appeared to be vanishing from much of their breeding range. Although the total global population of spoon-billed sandpipers can be difficult to gauge, estimates put the current number at no more than 340 individuals. One of the major threats to the species is habitat loss along their migration route; to travel from the Arctic to South Asia and back, they rely on a series of stopover sites along the coasts of China and the Korean peninsula, isolated patches of intertidal habitat spaced along the edge of the sea like a string of beads.

But with such a small number of birds migrating through such a vast area, it's difficult to know which potential stopover sites are the most crucial, and biologists and wildlife managers can't effectively target conservation efforts without knowing where, exactly, the birds are spending their time. That's where satellite tracking comes in.

When getting the project at APL off the ground, Bill Seegar had suggested bringing in the British scientist Paul Howey, whom he'd gotten to know during his research fellowship in England. Howey had built "bionic eggs" that transmitted data about the microenvironments in the nests of peregrine falcons, and Seegar thought his experience would be useful. Howey had also received a job offer at the University of Aberdeen to build transmitters for basking sharks, but the work happening at APL seemed as if it would be more of a challenge, so he came to America.

Howey had stayed on at APL to work on other projects after funding for the original bird transmitter project ran out, but he'd continued tinkering with bird transmitters on the side. The Argos system officially required transmitters to operate with at least one watt of power (an old-fashioned incandescent lightbulb, for comparison, is typically about sixty watts). When I interviewed Howey, however, he told me that he found during this after-hours experimenting that "if you got everything tuned exactly right so that the signal was absolutely perfect, you could go an order of magnitude lower than that, and it worked fine."

This meant he could make transmitters that were smaller and lighter. But because Howey's devices didn't meet their specifications, the French space agency initially refused to approve them for use on the Argos system. "They literally threw me out of their office, they said it won't work, go away, you're wasting our time," he told me. Undeterred, he had some biologist friends try his new gadgets out on eagles in the Grand Canyon anyway. They worked, the overseers of Argos gave in, and in 1991 Howey left APL to found Microwave Telemetry, a company that produces and sells transmitters for birds and marine animals.

By 2015, Paul Howey had gotten the weight of his transmitters down to just seven-hundredths of an ounce; that's about the weight of two paper clips, one one-hundredth the weight of the first bird transmitter prototype from the 1980s. He could produce only a small number of the tiny tags at a time, however ("You basically build them under a microscope," he said), and his company created an application process for bird biologists who thought they could put them to good use.

Spoon-billed sandpiper researchers leaped at the chance, but once they got approval to purchase a few of the transmitters, they had to be sure they would be safe and effective for use on the incredibly rare birds. Nigel Clark, a British ornithologist who chairs the U.K. spoon-billed sandpiper support group, explained the process they went through via a video call. The tags were originally designed to be attached with a harness, but Clark and his colleagues, concerned about how a harness might affect spoonies, wanted to use glue instead. To test this idea out, they first glued dummy tags to the backs of a captive flock of dunlins, a much more common cousin of spoon-billed sandpipers. Keeping the dunlins in an aviary in a neighbor's backyard, Clark was able to observe how, when, and why tags fell off in order to troubleshoot potential problems before they could endanger actual field work with spoonies.

By the fall of 2016, they were ready. Packing three of the precious transmitters—three chances to identify critical habitat and potentially save a species—Clark and his team traveled to the Yellow Sea coast of China's Jiangsu Province. The site they were headed for is not exactly the sort of pristine wilderness where you might expect to find one of the world's rarest shorebirds; behind the intertidal mudflats are seawalls protecting rice paddies and aquaculture ponds used to raise shrimp and fish. But it's one area where spoon-billed sandpipers on their way south from their breeding grounds in the Arctic pause each year to molt and grow new feathers before completing their migration.

In China, Clark worked with local ornithologists including the Hong Kong–born biologist Katherine Leung, who described for me how they caught their rare targets. Their most successful capture method involved stringing up mist nets in rice paddies at night, when the bamboo poles they used to hold up the nets were less visible. But although Leung and her colleagues were interested only in spoonies, "at those sites, there's a lot of other small shorebirds which will come in together with the spoon-billed sandpipers," Leung told me. "Each night we'd catch up to three hundred birds in those nets over several hours. We have to continuously extract the birds from the mist nets, otherwise they'll have problems with their wings or legs, so we have a team stay close to the net and check it and extract the birds every twenty minutes." Working in dark and sometimes wet conditions, they'd remove and release an onslaught of plovers and other more common cousins of the spoonies, all to catch perhaps three or four of their spatula-beaked targets.

Leung had been working with spoon-billed sandpipers at that site for years. But in 2016, she finally got to follow what happened to them after they left. "After we put satellite transmitters on in Jiangsu Province the first year, we were very excited to just keep checking the map every day to try to see where they would be going," she said.

Two of the three birds that Leung and Clark fitted with satellite transmitters that fall went on to other already-known stopover sites after they left Jiangsu Province. But the location fixes from one bird's transmitter showed it at a spot on the coast that no one had realized spoon-billed sandpipers were using. Clark and his colleagues were able to get in touch with local Chinese bird-watchers who, though skeptical, went to check it out. "They got to the site and found three spoon-billed sandpipers—not the satellite-tagged bird that day, actually, though it was seen later on—but they also found loads of nets up for hunting birds," Clark said. (Although spoonies are too small to eat, they are regularly captured and killed as bycatch by hunters hoping to net larger shorebirds to sell as food.) An associate of Clark's reached out to contacts in Beijing, "and by the next morning there were people on the ground, law enforcement, taking the nets down and putting up signs that there was to be no more netting."

Between 2016 and 2019, researchers put satellite tags on a total of thirteen spoonies, capturing some on their nesting grounds in Russia as well as at their stopover in China. The transmissions from the tags have helped them identify a total of twenty-eight stopover, molting, and "staging" sites (areas in Russia where the birds congregate to fuel up before embarking on migration). One newly identified molting site is located in North Korea, near the Korean demilitarized zone. The spoon-billed sandpiper task force is working on getting permission for a team to enter North Korea to do surveys there.

There's more work to be done. Only ten of the twenty-eight sites pinpointed by the transmitters are in areas that currently have any sort of legal protection. On top of that, scientists have yet to identify the Arctic breeding sites of 70 to 80 percent of spoonies' global population. And while Clark hopes to tag more birds, each transmitter costs five thousand dollars. "It's an expensive game," as he put it.

When Clark first got involved with spoon-billed sandpiper conservation, scientists estimated that the total number of spoonies in the world was decreasing by half every two years. "I was of the view that we were documenting extinction. We predicted that they would effectively go extinct in 2019," he said.

But this situation is changing. According to Clark, thanks to an enormous amount of conservation work that has been done throughout their range, the best estimate is now that spoon-billed sandpipers are declining at 8 or 9 percent per year, at which rate they will go extinct in the mid-2030s. "So, we have more to do," said Clark, "but we've bought time."

From Argos to GPS

Spoon-billed sandpipers may love intertidal marshes, but not all so-called shorebirds spend their time on the shore. Driving through a treeless prairie landscape in eastern Montana as dawn broke in May 2021, keeping an eye out for cattle, pronghorn, and sage grouse standing in the gravel road, I felt pretty far from the ocean.

The bird I was here to see was the long-billed curlew. Like spoonies, these shorebirds are known for their exceptional beaks. But while spoonies are small and delicate, the proportions of long-billed curlews are almost ridiculous. They look something like their cousin the godwit might if you took its bill and stretched it out as far as you could. The bill of a female long-billed curlew, curving gently downward near its tip, can be almost seven inches long. On a bird whose body is about two feet long, that's . . . a lot. I was really there to watch the Smithsonian Conservation Biology Institute ornithologist Andy Boyce fit them with backpack tracking devices, not just ogle their bills, but I'm getting ahead of myself: before I saw anything at all, we had to catch a bird.

It was the sort of day where even in the cool of the morning you can feel that it's going to be oppressively hot in a few hours. (Spring in Montana is strange; a few days later I would be caught in a snowstorm on my way home.) After meeting up with Boyce, the PhD student Paula Cimprich, and their crew of interns, I'd followed them to a site where they'd staked out the nest of the first curlew they hoped to capture. But something was amiss: neither parent was on the nest, even though one of them should have been incubating the eggs at all times. Boyce and Cimprich found a single intact egg, cold to the touch, and concluded that a fox or coyote must have gotten to the rest of the clutch in the night. The only easy way to catch breeding curlews is when they're hunkered down on their nests, so the chance to catch and tag that nest's female was gone, at least until she laid a new set of eggs.

On we went to nest number two, accompanied by an unending soundtrack of whistling, gurgling meadowlark song. Boyce and Cimprich shared notes on relocating the nest, well concealed in the grass—behind the upside-down tumbleweed, next to the big cow pie. Here, the female was on her eggs as she should be, and I got to watch from the road as they attempted to catch her using a technique called drop netting. Each holding one end of a mist net, Boyce and Cimprich crept up on the incubating female. Usually, the bird's instinct is to sit tight rather than flee. Not this time, however. She burst into the air when they were a few feet short, circling nearby and scolding until we departed.

Zero for two.

We had one more chance for that day. After driving to the next site, Boyce's interns and I made nervous small talk on the road as he and Cimprich set off into the grass with the net. We couldn't see the nest (when they returned, Cimprich said even she couldn't see it until she was almost on top of it), but we saw them suddenly drop the net, and then a large bird flapping underneath it.

Boyce carried the bird back to the pickup truck where his banding supplies were, a cloth hood covering her head to keep her calm. I had seen curlews at a distance before, but never so close. Patterned in subtle shades from buff to cinnamon, her wings and back barred with darker brown, and her eyes set off by pale face markings, she was a beautiful creature. With Cimprich's help, Boyce swiftly fastened a numbered metal band around her leg, measured her legs and wings, and briefly removed the hood to record the length of that fabulous bill. Then came the backpack. After slipping on the leg loops that would hold it in place between the curlew's wings, Boyce gently snugged them up, smoothed her feathers back into place, snipped off excess Teflon ribbon, and applied a drop of superglue to the cut ends so they wouldn't fray.

Unlike E7 or Leung's spoon-billed sandpipers, this curlew (whom Boyce would dub Frito, as in Frito-Lay, because he had once watched her lay an egg) was not wearing a device that transmitted a signal to an Argos satellite. She was being given a GPS logger, which would record her location with far more precision at roughly thirty-minute intervals. Barring unforeseen complications, she would carry it for the rest of her life as she flew from the Montana prairie to Mexico and back each year, giving scientists a window into every facet of her annual journey in the process.

Traditional satellite telemetry involves a device worn by a bird transmitting a signal to a satellite overhead. GPS (short for Global Positioning System) technology is, in a way, the reverse. Instead of sending information *to* a satellite, a GPS device receives information *from* satellites.

Developed in the 1970s by the U.S. military, the Global Positioning System consists of thirty-one satellites, each of which circles Earth twice a day at an altitude of around 12,500 miles. Every moment, at least four of these satellites are in the sky over almost every point on the planet. Aboard each one is an atomic clock, which uses the movement of electrons to keep time so accurate it may only be off by a millionth of a second every ten years. A receiver on Earth's surface can use the difference between the current time as transmitted by the satellite and the time the signal was received to figure out how far away the satellite is. If the receiver can pick up signals from four different satellites at once and knows the current location of each, it can put that information together to calculate and record its own current latitude, longitude, and altitude. GPS is so precise that it needs to account for Einstein's theory of relativity. As Einstein predicted, gravity and speed both have tiny effects on how quickly time passes at any point in the universe, and GPS needs to adjust for the resulting difference in time between orbiting satellites and earthbound receivers to prevent small errors that would add up over time into inaccurate positions.

For a period in the 1990s, as GPS technology advanced, the U.S. government intentionally tinkered with the system to make it less accurate for nonmilitary use (a practice they called "selective availability"). But just after midnight eastern daylight time on May 1, 2000, selective availability was switched off, a decision intended "to make GPS more attractive to civil and commercial users worldwide." The GPS signal accessible to civilians instantly became ten times more accurate, and ornithologists were on it immediately.

In the months before the end of selective availability, scientists in Germany were already experimenting with putting GPS devices on homing pigeons to record their flight paths over the countryside north of Frankfurt am Main. The location fixes provided by their devices were accurate to within 160 feet, but the researchers noted that the end of selective availability meant that future bird tracking projects would be able to record locations accurate to within 16 feet, an order of magnitude more precise than the Argos system. Within just a few years, GPS loggers were being used to follow the offshore foraging behavior of seabirds.

One problem with putting GPS trackers on birds is how to get the location data once it's been calculated and stored. The heaviest, most expensive GPS devices in use for birds today transmit their data via the Argos system even though they're not using it to determine their location, letting scientists access it at any time from anywhere in the world. The simplest and lightest devices, which can weigh as little as 0.03 ounces, don't transmit their data at all; ornithologists must recapture the birds wearing them and retrieve the devices themselves to download the information they carry. The GPS backpack I watched Andy Boyce put on Frito the curlew in Montana was designed to make use of two middle options. If the bird passes within range of a cell phone tower on the GSM network (the one that connects AT&T and

T-Mobile phones), it can transmit the data via the cell phone network. If that doesn't happen (not a given when the birds are on their breeding grounds in rural Montana, where, as I found, cell coverage is spotty at best), Boyce, Cimprich, and their colleagues can get within a certain range of a bird and download the data via a handheld radio antenna.

Boyce has put GPS tags on thirty-five curlews nesting on and around eastern Montana's American Prairie reserve since 2019. The birds that nest at his sites in Montana split up when they leave for their winter home: after a stopover in northern Texas and northeastern New Mexico, half head to the Gulf Coast of Texas and northern Mexico, while half spend the winter at five thousand feet or more above sea level in the high Chihuahuan desert of central Mexico. "They're experiencing two totally different habitats for 70 percent of the year," said Boyce, and issues in either of these two very different places could affect the breeding population in Montana.

Ornithologists had known about this split since 2014, when Lee Tibbitts was part of a group that published a study of curlew migration using traditional satellite telemetry. But GPS allows for data that's almost ludicrously detailed. Using an aerial image of the bird's desert wintering habitat in Mexico, Boyce showed me how the GPS points made it clear which individual irrigated fields the birds liked to forage in, data that allows his collaborators at the Mexican nonprofit Organización Vida Silvestre to better target conservation efforts there. On another occasion, someone working on curlew conservation in the Houston area learned about Boyce's work when they stumbled across a post about it on Twitter. They got in touch to ask if Boyce could help

---

* In 2019, a group of Russian ornithologists tracking eagles with GPS devices that transmitted their location via text message inadvertently racked up huge roaming charges when the birds failed to pass within range of cell towers while traversing Kazakhstan, instead sending a backlog of messages when they reached Iran, where SMS message rates are much higher.

them figure out where in the area curlews might be roosting at night. Boyce checked his data and advised them to take a look at the front lawn of the Houston Executive Airport that evening. Sure enough, there were the curlews—about a thousand of them.

When one tries to minimize the impact that a tracking device has on a bird, every fraction of an ounce matters. Boyce described starting with a tag that the manufacturer listed as weighing twelve grams (about four-tenths of an ounce). "But that doesn't include that the tag is white and you probably have to spray-paint it a darker color if you don't want your bird to be eaten by peregrine falcons," said Boyce. "It doesn't include the little neoprene pad that we put on the bottom of the tag to minimize chafing. It doesn't include the crimps to secure these things, which are half a gram each. So we're talking about more like sixteen grams before it actually gets on the bird."

Mindful of the potential for even the lightest, best-designed tags to have some unanticipated impact on the birds carrying them, he's also trying to wring every possible use out of the data he collects. While his collaborators in Mexico and Texas use his data to target conservation efforts on the curlews' wintering grounds, Boyce is using the data from the tags to study curlews' habitat preferences on their nesting grounds in Montana. They seem to like hanging out in and around prairie dog towns, but Boyce isn't yet sure why. One possibility is that prairie dog poop fertilizes the vegetation in a way that ultimately produces more of the beetles curlews love to eat.

And other ornithologists are interested in what the curlews are doing on their nearly twenty-five-hundred-mile journeys back and forth. Cimprich, a PhD student at the University of Oklahoma, is using the GPS data to study their in-flight behavior during migration— "How fast they're flying, what altitude they're flying at, and where they are along their route," as she explained it. By comparing this with local weather data from each area the birds pass through, she hopes to be able to determine how they use (or don't use) favorable winds to help them on their way. In addition to being interesting in its own right, this information can help scientists predict how curlews' migration might be affected by climate change, which is expected to increase the likelihood that the birds will face challenging headwinds during their fall migration. "They might have an extra energy expense, or maybe they're going to adjust by taking different routes," she said. "But we really don't know how they're going to react, because we don't have a good grasp of exactly what they're doing right now."

The fact that GPS tags—unlike traditional satellite telemetry through the Argos system—provide data on altitude as well as location is what makes them ideal for projects like Cimprich's. It also makes them a perfect tool for tracking another one of the world's most extreme migrations. Compared with the treks of the other birds featured in this chapter, the migration of bar-headed geese is just a short hop, perhaps less than a thousand miles, depending on where exactly an individual goose is headed. But in these geese's path is one of the most spectacular obstacles on the face of the planet: the Himalayan mountains.

Into Thin Air

Lucy Hawkes grew up dreaming of being a marine biologist, and she got her start in wildlife telemetry tracking sea turtles. But when she saw an advertisement for a postdoctoral research position to work with bar-headed geese in Mongolia, she decided to go for it. She couldn't quite figure out how to turn her interest in the movements of sea turtles into a career, "whereas birds, on the other hand, do these spectacular migrations," she told me when I caught her for a video call. "The fastest, the furthest, the longest, the highest—they're all birds."

Ornithologists knew that bar-headed geese spent part of the year on one side of the Himalayas, in China and Mongolia, and part of the year on the other side, on the Indian subcontinent. Mountaineers had reported seeing them at incredible-seeming elevations, including a member of Edmund Hillary and Tenzing Norgay's famous 1953 expedition to Mount Everest who claimed to have spotted a flock of geese flying over the mountain's twenty-nine-thousand-foot peak. But the details were sketchy, and that's where the plan to track them with GPS (using devices that would transmit the geese's locations via the Argos system) came in.

Hawkes's primary Mongolian collaborator was Nyambayar Batbayar. Growing up in Mongolia when it was still part of the U.S.S.R., he first got interested in birds when he joined a "young naturalists' club" overseen by the Communist Party. He'd been studying bar-headed geese since he was a graduate student, but the international team of researchers who converged on his home country for the satellite tracking study is something he remembers with enthusiasm. "The world's best experts in avian high-altitude physiology—they were in Mongolia!" he recalled when we spoke, finally connecting on a Sunday evening my time in order to work around our seventeen-hour time difference. "That was really something for me."

But just as with swans in Alaska, sandpipers in China, and curlews in Montana, the first problem in tracking the geese was figuring out how to capture them. "It was my fifth day on the job when I got on a plane to Mongolia," said Hawkes. "So far, I've worked with sea turtles on hot tropical beaches, and here we are on the plains of wild Mongolia, and there's this flock of six thousand bar-headed geese on this lake, and we have to work out how to catch them and no one has a clue what to do."

They were camping on the treeless plain surrounding Terkhiin Tsagaan Lake, a landscape Hawkes remembers as looking like something out of a watercolor painting. "The place was new for everybody except the Mongolians," said Batbayar. "It was the middle of July, and when we arrived at the site, the next morning, there was snow everywhere. It was something nobody expected to see in the middle of summer."

The one advantage the researchers had was that the geese were undergoing their annual molt and couldn't fly. Hawkes, Batbayar, and their colleagues eventually hit upon a strategy of gently herding the geese into nets on the shore using kayaks. When they captured a goose, they would use a harness to attach the tag that calculated the bird's location and altitude using GPS and transmitted the data via the Argos system. Once the geese finished growing in their new flight feathers and departed for India, there was nothing for the scientists to do but check their computers for updates from Argos and cross their fingers.

They found that the geese took only a day to cross over the Himalayas, some ascending nearly twenty thousand feet in about eight hours. The highest altitude at which the GPS devices recorded a goose flying was almost twenty-four thousand feet, but typically the geese weren't *literally* flying over the peak of Everest; they were following the terrain of the passes between the peaks. Surprisingly, however, the geese didn't seem to be making any effort to take advantage of favorable tailwinds. Like an airplane, Hawkes pointed out to me, they may get more lift by intentionally flying against the wind instead of with it.

One thing that makes such an extreme migration possible is birds' unique respiratory systems. Birds aren't just feathery mammals; their physiology is very different from ours. Like us, birds have lungs that exchange carbon dioxide in the blood for oxygen, but the similarities in how we breathe stop there. Bird lungs don't expand and contract; instead, air is pumped through them in a single direction by a system of air sacs. This is a much more efficient system than ours, and it helps explain how bar-headed geese can manage sustained flight through the thin air at the roof of the world.

After that first field season in 2008, Hawkes ended up returning to Mongolia every year for a total of seven years. "It was one of those experiences in life that challenged the heck out of me, and I got so much from conquering it," she said.

Humans, always interested in pushing the limits of our own physiology, have taken notice. One former colleague of Hawkes's, Jessica Meir, is now an astronaut; she applied for the astronaut program based on her expertise in the physiology of animals in extreme environments and went on to participate in the first all-female spacewalk in 2019. Hawkes was wearing a T-shirt commemorating Meir's space station expedition when we spoke, and she said she herself has been invited to speak at medical conferences by doctors interested in how lessons from bar-headed goose physiology could be applied to help people dealing with the many medical conditions in which the body struggles to get enough oxygen. "I don't see a translation of [what bar-headed geese do], other than turning your lungs into a unidirectional ventilated structure, which is never going to happen," Hawkes admitted. "But with COVID, there's been this thing called ECMO [extra-corporeal membrane oxygenation] some people have [as a treatment], where your blood gets pumped out of your body and gets oxygenated separately. That's kind of like a bird lung, in a way."

Satellites—first Argos, and then GPS—have let scientists like Lucy Hawkes, Nigel Clark, and Lee Tibbitts follow birds to places Bill Cochran could only dream of, and do it from the comfort and safety of their offices. But both of these systems were designed for other purposes before being co-opted by wildlife researchers. A new space-based tracking system, this one designed from the start for tracking the movements of birds and other animals, offers an example of what the future of wildlife telemetry may look like.

It's time to talk about Icarus.

The Internet of Animals

Martin Wikelski had been warned about Bill Cochran.

By the time the young German ornithologist arrived at the University of Illinois as an assistant professor in 1998, Cochran was no longer working for the Illinois Natural History Survey. Never interested in pulling his punches or playing politics in the way that can be necessary to advance in an organization over time, Cochran had apparently not left on good terms. (Cochran told me the trouble started when a profile of him in the *Chicago Tribune* failed to mention the name of the organization that employed him.) "People told me, if you ever meet this crazy dude, run as fast as you can," Wikelski said when we spoke via video call.

But one day when Wikelski was setting up mist nets in preparation for an attempt to capture and track Cochran's old subject the Swainson's thrush, Cochran—tipped off by a colleague of Wikelski's with whom he'd remained friends—showed up out of nowhere. It turned out "he's not a crazy person, he's just a fantastic engineer, a fantastic biologist, and a genius," said Wikelski. To a casual observer, the cantankerous seventy-year-old engineer and the ambitious young biologist from Germany might not have appeared to have much in common. But Wikelski shared Cochran's disregard for convention and his hunger to know where all those birds were going.

Wikelski left for a position at Princeton University after only two years in Illinois, but he and Cochran collaborated on a series of studies, investigating how thrushes recalibrate their internal magnetic compass, publishing some of the first evidence that bats navigate using Earth's magnetic field as well, and building devices to remotely monitor the heart rates of songbirds as they went about their daily routines. Wikelski also had an ongoing research project on Panama's Barro Colorado Island (BCI), an island in a man-made lake along the Panama Canal that has become one of the most studied patches of tropical rain forest in the world. He wanted to learn more about the interactions among the island's wildlife, and who better than Bill Cochran to help build a seven-tower automated radio telemetry system for BCI to track the movements of its animals? "Bill was soldering all the antennas in his garage in Illinois," said Wikelski. The system came online in 2003.

There was just one problem. BCI is so beloved by biologists because its seeming isolation makes it an ideal microcosm for investigating the rain forest ecology in detail. But one way or another, by air or by water, the animals that Wikelski and his colleagues tracked kept leaving the island. A tag from an ocelot turned up at the bottom of the lake after the animal that carried it was apparently eaten by a crocodile. A bee buzzed merrily off over the lake's surface to seek nectar on the mainland. "It was not an isolated place [after all]," said Wikelski, "but an open system." To get a full picture of the lives of

BCI's animals, they needed to be able to track them beyond the island's boundaries.

It was a visit by Cochran's old mentor George Swenson that inspired the next phase of Wikelski's career. Swenson, the Illinois radio astronomer with whom Cochran had designed radio beacons for satellites in the 1960s, visited BCI at Wikelski's invitation. Retired by then after a distinguished career at the University of Illinois, Swenson had gone on from his 1960s satellite work to lead the design of the Very Large Array, a network of twenty-eight enormous radio telescopes in the arid plains of New Mexico that had a starring role in the 1997 sci-fi film *Contact*. He was used to thinking big, and he immediately saw the solution to Wikelski's problem. Forget 130-foot towers; they needed receivers in space, and the Argos system wouldn't cut it. At the turn of the millennium, 80 percent of the world's bird species were still too small to track using existing satellite technology, not to mention two-thirds of the world's mammals.

Officially, Wikelski called the system he ultimately dreamed up Icarus, for "International Cooperation for Animal Research Using Space." Informally, though, he loves to refer to it as "the internet of animals," inspired by the concept of the "internet of things," the billions of everyday physical objects around the world that are embedded with sensors and software to collect and share data via the internet. Icarus would rely on GPS for location fixes. But its receiver would be in a lower orbit than the Argos satellites, making smaller transmitters possible. Crucially, the system would also encode transmissions digitally, allowing many more tags to operate simultaneously. By assigning each transmission its own unique code, the protocol used by Icarus lets many transmitters share the same radio frequency without confusing the receiver. The Argos system, with its 1970s-era technology, does not do this.

Wikelski originally conceived of the idea that would become Icarus in 2001. Making it a reality would be the focus of the next two decades of his career. He pitched the idea to NASA in 2004, but they turned him down, just as they had turned down Cochran's original satellite-tracking scheme forty years earlier. In 2008, Wikelski moved back to Germany to take over the Max Planck Institute for Ornithology, where he became a controversial figure for his single-minded pursuit of Icarus; he even shut down the institute's long-running bird banding program.

Finally, in 2012, the German and Russian space agencies agreed to provide funding and space on the International Space Station. But Wikelski's wait wasn't over: a planned 2015 launch was pushed back to 2018 by a flare-up of tensions between Russia and the West. A pair of cosmonauts finally installed the Icarus antenna on the exterior of the space station during a 2018 spacewalk, only for a malfunction in the device's onboard computer to necessitate another delay for repairs. After the fix was completed, the system was supposed to be available to scientists beginning in the fall of 2020, but the first transmitters shipped out had a defect that allowed water to leak into the casing. "The space engineers always told us, it's gonna take four years longer than you expect," said Wikelski.

The future of Icarus is still not assured; in the spring of 2022, the system was shut down by the war between Russia and Ukraine, when a Russian ground station through which data was being relayed to the International Space Station abruptly stopped transmitting. But as I was first researching this in the summer of 2021, data was rolling in from animals around the world—not just birds, but tortoises, bats, rhinoceroses, and more. On Movebank, the free online database where Icarus data is archived, I was able to pull up a map showing the journey of a Hudsonian godwit (close cousin of the Pacific-traversing bar-tailed godwit) that had been fitted with an Icarus tag on its wintering grounds in Chile. After a long open-water flight north over the eastern Pacific, it had touched down in Mexico, crossed the Gulf of Mexico, and eventually arrived in the Canadian prairies on May 17.

The first generation of Icarus transmitters weigh just under a fifth of an ounce and cost €500 each, or about $590 (the closest equivalents available for the Argos system cost five times as much). In the coming years, Wikelski plans to get the weight down to less than one-twentieth of an ounce, as well as launch independent Icarus satellites to eliminate the system's reliance on the International Space Station (and, by extension, on Russian relay stations). And it's worth noting that Argos isn't taking this lying down. The arm of the French space agency that oversees the Argos system is preparing to launch a new constellation of twenty-five tiny "nanosats" that will allow for more frequent location fixes and open new possibilities for scientists using the system to customize and communicate with their transmitters.

I emailed Juan Navedo, the Chilean ornithologist who tagged the godwits there, to ask him what he wanted to learn from his Icarus transmitters. The lower cost per transmitter, he replied, means he can tag more birds than his budget would otherwise allow. "By deploying a number of tags, we will increase the chances of finding 'hidden movements' of migratory animals," he wrote. "This will increase our chances to explore individual decisions and identify additional important [areas] for their conservation."

In the Anthropocene, even birds capable of the most extreme physiological feats will need human help to survive into the future. And if we don't know where birds are going, we don't have the information we need to save them. Luckily, we have eyes in the sky.

Six

Navigating by the Sun

The migratory songbirds most beloved by North American bird-watchers are, arguably, the wood warblers. Of the hundred-plus species in this family, roughly fifty breed in North America and migrate south for the winter, many crossing the Gulf of Mexico to spend the winter in Central or South America. Although some species are colored in subtle browns and grays, the plumage of many male warblers includes splashes of intense color—the glowing orange throat of the Blackburnian warbler, the flashes of red in the wings and tail of the American redstart, the deep sky-blue back of the cerulean warbler, the yellow cheeks, throats, wings, and rumps of many other species. Mostly insect eaters, warblers seem to be in perpetual motion, flitting from perch to perch with restless energy as they hunt for tiny bugs and caterpillars.

Despite their bright colors, warblers can be hard to spot if you're not actively looking for them. They're tiny, and many species prefer to spend their time high in the treetops or concealed in dense brush. But that just makes them all the more beguiling to birders. Finding one can be like finding a hidden gem.

Habitat loss threatens nearly a quarter of warbler species. But as with all migratory birds, if we don't know exactly where they're going, it can be difficult or impossible to effectively target conservation efforts; even the most stringent efforts to protect a warbler's breeding habitat in North America won't stop a population from declining if the specific patch of tropical forest it relies on during the winter is being cut down. Being able to track these birds from North America to the Neotropics and back would provide crucial information for wildlife managers working to save them.

Depending on the specific species, however, a warbler may weigh half an ounce or less. The smallest Argos transmitters available from Paul Howey's company, Microwave Telemetry, weigh 2 grams, or 0.07 ounces. The general rule is that you shouldn't use a transmitter that's more than 5 percent of a bird's body weight, so the smallest bird you could put one of these on would weigh about 1.4 ounces. Spoon-billed sandpipers are right at that edge. Warblers? Not a chance.

The smallest, lightest GPS devices are called loggers, because they store location data on board but don't transmit it. This saves weight, but it also means you have to recapture the bird, remove the GPS logger, and physically download the data from it—plus they can record only a limited number of location fixes. They weigh about a gram, or 0.035 ounces. According to the math, the smallest bird that could wear a GPS logger would weigh 0.7 ounces. You could safely put one on, say, a house finch, those little brown birds with red faces that visit backyard feeders across North America. But not a warbler.

The "nanotags" used by the Motus automated radio telemetry system can be tiny. Really, really tiny, as small as 0.26 grams or a hundredth of an ounce. Tags for automated radio telemetry can be so small that researchers have even experimented with putting them on hummingbirds. But if you're going to use one of these tags to study warbler migration, you'll have to cross your fingers that your bird will regularly pass within range of Motus towers all along its route. If the whole reason you're doing your study is that you're not sure where your warbler is headed, that could be a risky proposition.

There is a solution, however. The answer lies in a method of navigation used by seafarers for centuries, long before the development of modern technology like GPS satellites. All you really need to keep track of where in the world you are is an accurate clock—and the movement of the sun.

"This Will Never Be Useful"

The British biologist Rory Wilson wasn't thinking of warblers when he first conceived of what came to be known as light-level geolocation. He was thinking of penguins.

In the late 1980s, Wilson was a researcher at Germany's Institut für Meereskunde studying the movements of penguins, which, during the breeding season, alternate with their mates between shifts incubating their eggs and long foraging trips into the ocean. The long-distance transmitters that existed at the time were big, heavy, and awkward, and he was obsessed with coming up with a better way to figure out where his penguins went when they left their nests behind. One of his first attempts at a solution involved "dead reckoning": devices that recorded penguins' speed and trajectory at regular intervals, reconstructing their movements from there. It worked at small scales, but if the birds traveled too far, the influence of ocean currents began to cause errors.

Some accounts claim that Wilson's moment of inspiration for what became known as light-level geolocation came directly from historical mariners. "Wilson recalled that early seafarers didn't have onboard GPS, nor satellite transmitters to ascertain their position: they observed the sun's trajectory to estimate their whereabouts," wrote one of his former students. But when I reached Wilson by video call, he admitted that he'd only recognized the historical parallel after the fact, having first hit upon the idea while pondering what he could do with the technology he had access to. "I was playing around with magnetic field, declination, and the rest of it, but at the time, sensors for measuring that stuff just weren't available," he told me. Sensors that could detect sunlight, recording when the sun rose and set—those he could get. And they were relatively lightweight.

Today, Wilson is a professor at Swansea University. He's appeared in BBC wildlife documentaries and been hailed as one of the U.K.'s "top conservation heroes" alongside familiar figures such as David Attenborough and Jane Goodall. But he made time to chat with me about the early days of light-level geolocation on a late Friday afternoon U.K. time, sitting in his home office dressed in an oversized T-shirt as he squeezed our conversation in before starting the weekend.

The length of a day varies with latitude—how near you are to the equator or the poles—and the time at which the sun is highest in the sky changes as you travel east or west. The basic idea that observations of the sun's position in the sky could be combined with accurate time-keeping to estimate one's location on the surface of the planet dates to the 1530s, but it would take another couple centuries for anyone to develop a clock precise enough for seafarers to use in this way. Wilson's idea relied on these very old principles. His devices consisted of nothing more than a light sensor, a clock, a memory chip, and a battery.

The ornithology community was skeptical. "Starting around 1995, I tried to publish stuff," he said. "I wish that I'd kept the rejection letters, because there were things like this will never be useful, the best-case scenario is that it will enable you to define which ocean basin the bird might be in."

Wilson's first published study using what he called Global Location Sensors, or GLSs, followed the foraging trips of Magellanic penguins breeding on the coast of Argentina. Ten birds were fitted with the 1.4-ounce devices, which recorded ambient light levels roughly every minute and a half. After ten weeks, Wilson and his colleagues recaptured all ten penguins to see where they'd been. The location fixes provided by the GLSs were only accurate to within about 25 miles, but that was enough to determine that they'd traveled as far as 180 miles from their nests, many heading to an area of ocean east of the colony where the small fish they liked to eat must have been especially dense.

The need to physically recover the devices to download the data did sometimes pose problems. At one point Wilson gave a few GLSs to a colleague studying chinstrap penguins that bred on an island off the coast of Antarctica. The penguins left for the winter, and when they returned to their colony the following spring, Wilson's colleague called to tell him he'd spotted a couple birds with the devices on their backs. "I said, for God's sake, pull them off! They're not stable for very long, the feathers wear off! And he said, oh, I'll go in the next few days," said Wilson. "And in a few days, he went and there was only one left. The other birds that had had them had sort of a little patch of feathers missing where the tag had been, but he couldn't find the tags. So what should have been a study of the overwintering movements of five chinstrap penguins became a study of the overwintering movements of *a* chinstrap penguin."

Wilson spent years refining the technology. One major advance was the discovery that adding a blue filter over the light sensor made the device less sensitive to disruptions from cloud cover, because the blue end of the spectrum is least likely to be filtered out by clouds. But by the turn of the millennium, he'd largely stopped working on light-level geolocation, discouraged by lack of interest from others in the ornithology community.

The story might have ended there if not for a Russian electrical engineer, Vsevolod Afanasyev, who worked for the British Antarctic Survey (BAS). Various sources describe Afanasyev as "extravagant," a "genius," "idiosyncratic," and "basically an inventor manqué." (I confess I had to look up "manqué," but apparently it means someone frustrated in their ambitions; that is, Afanasyev really wanted to be a famous inventor and not just an engineer assembling gadgets for British scientists.) A journalist who joined an expedition on a BAS research ship described how at night Afanasyev "would prowl silently about the vessel, unmistakable with his long grey beard and glasses as thick as his Russian accent," spinning "utterly absorbing Russian tales" for anyone who would listen. He was, in short, a character.

According to an account by the ornithologist David Grémillet, in the early years of the twenty-first century Afanasyev "was getting bored in the labs of the British Antarctic Survey in Cambridge. His bosses, previously some of the fiercest critics of geolocation, did not really know what to do with him and suggested 'why don't you build us a super mini light-recording device?' To their astonishment Afanasyev succeeded, and his neat 9 g [0.3 ounces] waterproof tag, capable of recording light levels for several years, was presented to baffled scientists at the first international biologging conference in Tokyo, in the spring of 2004."

Afanasyev's device weighed less than a quarter of the ones Wilson had been working with, and BAS quickly realized its potential value. Soon BAS scientists were not only conducting large-scale studies of albatross movements using light-level geolocators but also producing extra devices to share with collaborators at other organizations. Geolocators were small enough that when used on large seabirds, they could even be built into leg bands. This eliminated the need for harnesses, to which seabirds are especially sensitive.

When I first started researching the history of geolocators, I hoped to interview Afanasyev, for obvious reasons. Unfortunately, after a fruitless search for current contact information for him, I learned that he had passed away in 2018.

Today private companies have largely taken over the task of producing light-level geolocators for bird research. Thanks to the fact that the components needed to build them are so simple, there are now light-level geolocators available that weigh as little as 0.013 ounces. (After casting about for an example to illustrate how light that is, I determined that it's very roughly the weight of one shelled pumpkin seed.) As of May 2022, a Google Scholar search for "light-level geolocator" yielded around twenty-two hundred results.

Rory Wilson doesn't seem to bear Afanasyev any ill will; the two never met, and Wilson had already taken the technology as far as he could before Afanasyev's breakthrough. But when I spoke to Wilson, I asked if he'd ever envisioned such a future when laboring away in obscurity over his penguin trackers.

"To think that it was going to become as small as it's now become, and become as potent as it's now become . . ." He trailed off for a moment. "In my wildest dreams, I wouldn't have thought that."

Out to Sea

As geolocators became smaller and smaller, they offered a new window into the migrations of songbirds whose routes ornithologists had previously been forced to guess at with only limited data. The ornithologist Ian Nisbet, whom you might remember from the first chapter of this book, used banding records from the eastern United States to assemble a theory about bird migration that seemed outlandish at the time in the late 1960s. Blackpoll warblers spend the summer in the boreal forests of northern North America and the winter in northern South America. In spring, they cross from South America to Florida before fanning out across North America. But during fall migration, Nisbet noted, banding records showed that blackpoll warblers were curiously absent from the southeastern United States. He suggested that instead, they were embarking from New England in the fall and reaching South America via a long nonstop flight over the western Atlantic.

This may seem almost trivial in comparison with E7's flight across the Pacific. But remember, a bar-tailed godwit is roughly the size of an American football. Female godwits preparing for migration can weigh almost a pound. The very fattest premigration blackpoll warblers, on the other hand, weigh about half an ounce, about equivalent to an AAA battery. These are very small birds. And Nisbet was proposing that they fly well over a thousand miles over the open ocean.

Many in the ornithology community agreed that this was the only possible explanation for the available data, but not *everyone* thought it was plausible. One ornithologist in particular, Bertram Murray, published lengthy rebuttals of Nisbet's theory well into the late 1980s, in a scientific back-and-forth reminiscent of the one between George Lowery and George Williams about whether birds really migrated across the Gulf of Mexico.

It was light-level geolocators that finally settled the matter.

Bill DeLuca spent his PhD years at the University of Massachusetts studying the ecology of blackpoll warblers and other birds that shared their habitat in the mountains of New Hampshire, surveying birds from the hiking trails of the White Mountain National Forest. He finished in 2013 and stayed on at the university as a research fellow, still unsatisfied with how little anyone really knew about these warblers' lives. Light-level geolocators had been getting smaller and smaller for years, first used on wood warblers in 2009 for a study of the ovenbird, a relatively large member of the family. A former mentor of DeLuca's suggested that he try them out on blackpolls.

Blackpoll warblers' name comes from the black "cap" on the tops of their heads. Foraging in conifers for tiny insects, nesting blackpoll warblers are widespread in Alaska and the Yukon, but in New England the type of boreal forest they prefer is found only on the tops of mountains. An organization called the Vermont Center for Ecostudies had been conducting a long-term mist netting and bird banding project on top of Mount Mansfield, the highest peak in Vermont, and DeLuca partnered with them for his field work in the summer of 2013.

"Typically that habitat is difficult to access, but [Mount Mansfield] is part of Stowe Mountain Ski Resort, so there's a road that gets you to the top of the mountain," he told me when we spoke. "It was kind of fun. We would drive up there and set up twenty or so nets along the trail system up there. We'd open the nets for an evening banding session, maybe from four until sundown, and we'd catch as many birds as we could and put the transmitters on. Then we would sleep in the ski patrol hut on the top of the mountain and wake up at 3:30 in the morning and do it all over again. We did that once a week for several weeks."

They put geolocators on twenty blackpoll warblers that summer. "It was a total pain in the butt to put the geolocators on [the birds] the way we did it for that project," DeLuca admitted, though warbler researchers have improved and standardized their techniques in subsequent years. "We used fluorescent-orange fly-fishing line as the harness material, because it was really strong, but the problem was you actually had to tie it on the bird and it was really hard to get it situated correctly."

Of course, putting the geolocators *on* the birds was only the first half of the challenge. The next year, DeLuca and his colleagues had to recapture those same birds when they returned to Vermont after their winter in the tropics—or at least they had to attempt to. "We would run around with binoculars and look for banded birds and use playbacks [of their calls] and decoys to get them back into the nets," said DeLuca. Because nothing quite like this had been done before, he would have been delighted to recapture even one. Ultimately, they were able to recover three of their devices.

Fortuitously, they learned that another group of researchers had put geolocators on blackpolls in Nova Scotia the same summer that DeLuca had been tagging birds in Vermont. That group also put out around twenty devices and had managed to recapture two of their birds. The two teams decided to pool their data for a single paper.

Reviewing the data from a light-level geolocator is not as simple as plugging it into a computer and instantly seeing exactly where the bird has been. Assuming the tiny device is still running, you use specialized clips to connect it to a computer and download the raw data, which is just a string of numbers representing dates and times and light measurements, requiring several rounds of analysis to make sense of. If the geolocator's battery is dead, the process takes even longer, because you have to send it back to the manufacturer and have them retrieve the data.

"There wasn't necessarily one eureka moment where you see the map and you're like, oh my God, there it is," said DeLuca. There was lots of talking on the phone, lots of emailing data and interpretation back and forth between the research groups. Eventually, however, the results became clear. Between September 25 and October 21, four of the five birds whose geolocators had been recovered set out from a stretch of Atlantic coast between Nova Scotia and New Jersey. (The fifth continued to follow the coast south before departing from Cape

Hatteras, North Carolina, on November 4.) From there, each bird had made a long flight over the open waters of the Atlantic to reach Puerto Rico, Hispaniola, or Turks and Caicos, pausing for a stopover on one of those islands before continuing on to northern South America, where they spent the winter. The longest single transoceanic flight recorded by the geolocators was fourteen hundred miles, completed in about three days. For three days and three nights, a bird that could sit in the palm of your hand beat its wings without ceasing, with nothing below but the endless blue and gray and green of the ocean.

I asked DeLuca *why* blackpoll warblers do this, but he said there are no concrete answers, only guesses. "There's a good amount of evidence that migration is the riskiest time of year" for migratory songbirds, he said, "so it really benefits them to just get it over with as quickly as they possibly can." In the fall, when weather patterns are right to make such a long overwater flight possible, it lets them avoid the predators and other hazards they might encounter flying over land.

Like E7's amazing feat the decade before, the extreme migration of the blackpolls made a media splash when the results were published in 2015. "I'd had papers before that had interest from the media, but I wasn't thinking I was gonna be on *Science Friday* and in *The Boston Globe*," said DeLuca, now a migration ecologist with the Audubon Society.

In 2019, DeLuca and his colleagues published a follow-up study that tracked the full migration of blackpoll warblers that bred in Alaska, documenting an incredible 12,400-mile round-trip journey in which birds first crossed the full breadth of the continent from west to east before heading out across the ocean. According to the geolocator data, some of the Alaska birds *did* actually travel all the way south to Florida before making their crossing in the fall, instead of setting out for a superlong flight from New England, and Ian Nisbet himself co-authored a response arguing that that must have been a mistake, once again citing the lack of banding records from the area. Whoever is right, one thing is clear: in miles traveled per ounce of weight, the blackpoll warbler is one of the world's greatest travelers.

Into the Swamp

I had seen ornithologists put various tracking gadgets *on* birds, but even after talking to DeLuca, I was having trouble picturing how one would go about recapturing the exact same tiny bird in a dense expanse of forest a year later, after it had flown to another continent and back, to retrieve a geolocator and download the data. If there was ever an appropriate time to trot out the cliché about needles and haystacks, surely this would be it.

It was this curiosity that led me to a parking lot on the campus of Louisiana State University, George Lowery's old stomping grounds, at six o'clock one morning in late April. Although the sun wasn't up yet, it was already so humid that the windows of my rental car had fogged up and my shirt was sticking to my skin. The predawn air was filled with the car alarm songs of northern cardinals. It had been a while since I'd been to a part of the country with cardinals, and I'd forgotten how *loud* they could be.

I wasn't there to see cardinals, though. I was there to meet up with the LSU graduate student Garrett Rhyne, who was studying a little-known oddball of the warbler family, the Swainson's warbler. Patterned in brown and soft yellow instead of the flashier colors of some of their cousins and sporting the longest beaks of any warbler, these birds love dense, swampy thickets, and this preference has given them a reputation for being hard for bird-watchers to spot and difficult for ornithologists to study. (George Lowery's backyard backed up onto a swamp, and apparently his property became known among LSU bird-watchers as a reliable place to see Swainson's warblers.) The previous spring, however, Rhyne and colleagues in other parts of the Southeast had outfitted eighty-seven of these little birds with light-level geolocators. Now, the real challenge had arrived. It was time to recapture them.

From the parking lot, I followed Rhyne to a plot of land north of town as daylight grew. The property owner, Dorothy Prowell, herself a retired LSU professor, met us at the gate at the end of her driveway. Prowell, her sister, and I were soon trailing Rhyne along the edge of a cypress-lined pond and into the woods, Rhyne carrying the metal poles for his mist net over his shoulder.

As we walked, Rhyne told me more about why he was studying Swainson's warblers and what he was hoping to learn (with a brief interruption from Prowell to warn me to keep an eye out for venomous snakes). "There are three populations," he explained, one along the East Coast, one along the Gulf Coast, and one farther north in the Appalachians, which is where he first fell in love with the species as an undergraduate. A small pilot study with just a few geolocators, done by a previous student in his lab at LSU, had suggested that the southeasternmost birds wintered in Cuba, Jamaica, and the Bahamas, while the more western birds, the ones nesting where we were, headed for Guatemala and the Yucatán. The Appalachian birds' migration patterns remained a total mystery, but Rhyne was hoping to change that.

Swainson's warblers can be hard to find because of their specific habitat needs, but "it's hard to say whether they're declining," according to Rhyne. "I don't think Breeding Bird Surveys pick them up accurately, because those are roadside surveys and you really have to get off road [to find them]. So we just don't know." (Breeding Bird Surveys are one of the main tools for estimating the population trends of North American birds; we'll return to them in chapter 9.)

We walked until Rhyne's handheld GPS unit told him we'd reached the spot where he'd captured a bird and deployed a geolocator a year before. There, the four of us stood and listened for any hint of singing Swainson's warblers; I wouldn't have been able to pick them out from the din of other songbirds, but my three companions knew exactly what they were listening for. They didn't hear it, apparently, but Rhyne retrieved a small speaker from his backpack and pulled up a recording of the song. If the male bird was around, the sound of an intruder on its territory should draw it out.

"There he is!" It worked. The male flitted into view at eye level, just off the trail, then briefly zipped out toward us before vanishing again. "Wow, you see how aggressive they are," said Rhyne. Managing to spot the bird again in his binoculars, he confirmed that it was wearing colored leg bands and a tiny backpack. Wherever it had been over the winter, Rhyne's bird was back.

After that, catching it was almost comically easy. Rhyne pulled a mist net from his backpack, drove the metal poles into the ground, and strung it out along the trail, leaving the speaker propped up nearby. We retreated a short distance away, and the male, incensed by the ongoing sounds of what he thought was a singing competitor, bounced into the net in about thirty seconds. (Females are much more difficult to catch; all of the birds in Rhyne's study were males.) If I'd been faster with my camera, I could have gotten a photo of Rhyne raising his eyes skyward and pumping a fist in triumph before untangling the bird from the net and placing it in a cloth bag.

He had a whole mini bird banding station in his backpack, and spreading out his equipment on the trail, he sat down and proceeded to measure and weigh the bird and take a blood sample. First, though, he snipped through the loops of jewelry cord that held the geolocator in place on the bird's back, gripping the bird securely but gently with one hand and working at the device with the other. "Ahh, relief," he said to the bird as it came off. (I could only imagine it must feel a little bit like taking one's bra off at the end of a long day.) Finally, Rhyne placed the geolocator in a small Tupperware container and sealed the lid.

"A safe spot for you," he said. "My master's degree, riding on that." Over the rest of that morning and the next, I would see that it wasn't always so easy: mortality on migration is high, and many of the spots where he'd captured birds the previous spring were now empty of Swainson's warblers, or had been taken over by different birds from the ones he'd caught last year. (He had also given some birds identifying bands without putting geolocators on them, to check whether the tracking devices themselves affected the likelihood of the birds reappearing.)

But with Swainson's warblers, at least, recapturing the birds that *did* turn up again appeared surprisingly simple. These birds would soon reveal whether and to what degree the three populations of

Swainson's warblers showed what ornithologists call "migratory con-nectivity," a pattern where birds from a specific breeding population within a species stick together and migrate to a specific wintering lo-cation, instead of scattering across the species' full winter range. For many species—not just warblers, but all sorts of birds—getting to this level of detail may be a key part of identifying and responding to the distinct threats that are causing them to decline.

Connecting the Migration Dots

Part of the inspiration for Garrett Rhyne's Swainson's warbler study came from a talk on geolocators he'd seen by a researcher named Gunnar Kramer years before. Kramer was "one of the smartest people I've ever met," said Rhyne, "and I was scared to walk up to him as an undergrad. I was like, hey, awesome presentation, I'm studying Swain-son's warblers, and I think this would be really cool [to do with them]. And he really pumped me up, he said yes, if I could do another one of these projects on any other bird, it would be Swainson's warblers. You need to do that."

Few people have spent more time tagging warblers with geolo-cators and thinking about the resulting data than Kramer, currently a postdoctoral research fellow at Harvard University. If anyone could give me the big picture of how geolocators are improving our under-standing of these birds, it was him, and I'd arranged to speak with him via video call before my trip to Louisiana. By his own admission, he's lost count of the exact number of warbler species he's done geolocator studies with, though he guesses it's up to ten or twelve now.

Kramer's ornithology origin story is one I've heard many times now: He was studying science as an undergraduate and considering a career in medicine, but felt uninspired. On a whim, he decided to take an ornithology class. He fell in love with birds and never looked back.

Eventually, he got involved with research on golden-winged war-blers, a species with dapper black facial markings and splashes of bright yellow on its wings and the crown of its head. Golden-winged warblers breed in the eastern United States and have been declining in many parts of their range. These declines can be partly blamed on competition from and hybridization with their close cousin, the blue-winged warbler. But why birds in the Great Lakes region seemed to be holding their own even as populations in other regions, such as the Appalachian Mountains, were dropping sharply was a mystery.

As a master's degree student in Minnesota, Kramer used radio transmitters to track the local movements of fledgling golden-winged warblers after they left their nests. Working with the University of Toledo ornithologist Henry Streby, under whom he would eventually study for his PhD, Kramer first tagged golden-winged warblers with geolocators in 2013—the same year Bill DeLuca deployed geolocators on blackpoll warblers. Yet another researcher tried them out on prothonotary warblers that spring. Tracking devices small enough for these iconic Neotropical migrants had finally arrived, and ornithologists were eager to take advantage of them.

Eventually, Kramer and his colleagues would recover more than forty geolocators from both golden-winged warblers and blue-winged warblers across their breeding range. Their data showed that instead of mingling indiscriminately when they arrived in Central and South America, golden-winged warblers that bred in specific areas of North America also spent the winter together in specific areas within their nonbreeding range: birds that bred in Minnesota all went to southern Mexico and Nicaragua, birds that bred in Tennessee all went to the border of Venezuela and Colombia, and so on. In other words, they demonstrated high migratory connectivity.

"It provided a mechanism to explain this fifty-year-old mystery of why golden-winged warblers were declining," said Kramer, "and why all the actions that we were doing to create new habitat for them on the breeding grounds failed to explain those really starkly different population trends." If all of Appalachia's golden-winged warblers were converging on a specific forest in South America that was being destroyed by logging, for example, then no amount of effort in their

North American range could keep them safe. Migratory bird conservation requires coordination on a global scale.

How many geolocator-tagged birds will be recaptured is often the biggest question mark when planning a project, and according to Kramer different species and even different individual birds vary in how easy they are to catch. "Some birds are just net happy," he said, like Garrett Rhyne's Swainson's warblers. Other birds, however, are wilier. "I remember a hybrid golden-winged/blue-winged warbler in Michigan where me and two technicians spent seven days just focused on this one bird, just packing nets into this bird's territory," said Kramer. "On the last day, our last chance, he barely caught the top of a net and fell in. It was pure luck."

The surprises never quite seem to end when it comes to bird migration. Just two years after Bill DeLuca published his study confirming the long-predicted transoceanic flights of blackpoll warblers, news broke that a *second* warbler species was doing the same thing—this time one that no one had really been paying attention to. Connecticut warblers are relatively plain looking by warbler standards, clad in a subtle blend of gray, olive brown, and yellow. Skulking in boggy thickets, they can be hard to spot, and although ornithologists knew that they, too, wintered in South America, for a long time exactly where they went or how they got there was a mystery. Banding records during migration were sparse, but one possible clue came from a cruise ship in the Caribbean whose lights attracted a whole flock of migrating Connecticut warblers in the fall of 2002.

The University of Manitoba ornithologist Emily McKinnon, who had studied under the first researcher to use geolocators on songbirds, became interested in Connecticut warblers after moving to a region of Canada where they were common during the breeding season. "I was doing a postdoc on snow buntings, but I thought, well, I can drive an hour from my house and be in a bog where there's Connecticut warblers," she told me when I interviewed her about the study.

Fitting the project in around her snow bunting research, she tagged twenty-nine Connecticut warblers with geolocators in the spring of 2015, recaptured four the next year, and confirmed that in the fall they made nonstop two-day flights over the Atlantic to reach stopover sites in Cuba and Hispaniola from the East Coast of the United States, following a route similar to that of blackpolls. (No knotting orange fishing line for her, however; McKinnon's former advisor, Bridget Stutchbury, was an avid needlepointer who'd taught her how to individually stitch each tiny harness into place.) McKinnon thinks there are probably more small songbird species that make this trip but just haven't been tracked yet.

Geolocators will never be able to match the pinpoint accuracy of GPS. Estimates of how far off the average location calculated from geolocator data is range from seven miles to more than three hundred miles, depending on the type of study and the method of analysis. They tend to be more accurate on the east–west axis than north–south, because calculating a precise latitude depends on recording accurate sunrise and sunset times, which can easily be thrown off if the bird is in a shady environment. But even that can be enough to give ornithologists the first indication of where in Central or South America warblers from a specific population are spending the winter, and in addition to being very small, light-level geolocators are less expensive than other types of tracking devices for birds and can record data every two minutes for an entire year. "That's a lot of information," said Kramer, and "it's helped us answer a lot of questions about the ecology of these species."

Kramer and McKinnon both remain fascinated by the continent-spanning journeys of these tiny birds. Migration is "one of the coolest, most interesting phenomena that exists on our planet," said Kramer. "Migration is just this crazy thing, and it's difficult not to anthropomorphize, to think about how challenging migration must be for these birds, when in reality it's what they're built to do."

But in addition to simply being amazing in its own right, the data coming from geolocator studies is essential for fully understanding the lives of warblers and, as a result, the reasons behind the ongoing declines of many species. As Kramer's data on golden-winged warblers showed, conservation efforts for birds that breed in North America aren't sufficient if they focus primarily on threats that the birds face *in* North America. After all, migratory species may spend more than half the year elsewhere. "It's like if you're trying to eat healthy, but you're only focusing on your breakfast, and you're not paying attention to what you're eating for lunch or dinner," said McKinnon. "If you only have this small window where the bird is on the landscape [that you're studying], it's pretty hard to assess what the major threats even are. With geolocator data we can start to look at the bigger patterns." This sort of tracking data can help persuade North American bird lovers to support conservation in places that may otherwise feel very far away.

Warblers are far from the only birds that have been tracked via geolocator. The first migratory songbirds to be tracked this way were the wood thrushes and purple martins tagged by McKinnon's PhD advisor, Bridget Stutchbury, in 2007. In Europe, geolocators revealed for the first time that the red-necked phalaropes that breed in Scotland travel halfway around the world to spend the winter off the coast of Ecuador and are being used to study the unusual east–west migration route of common rosefinches. But there's just something special about warblers. There are so many of them, they're so tiny and so brightly colored, and they travel such long distances.

But how small does a bird have to be, and how far does it have to fly, before carrying a tracking device on its back—even one as lightweight as a light-level geolocator—becomes a burden?

Balancing the Scales

Light-level geolocators are a marvelous fusion of modern technology and ancient navigation principles. They've helped uncover crucial information about the migratory routes of threatened birds. But they aren't perfect.

One obvious downside of a tracking device that requires you to recapture the bird and download the data is that you never learn what happens to the birds that don't make it back. A 2021 study of geolocator-tagged painted buntings, colorful songbirds that breed in the southern and southeastern United States, found that buntings that spent the winter in Cuba were 20 percent less likely to return the following spring than birds that wintered in other parts of the Caribbean. If researchers weren't careful in interpreting their data, they could easily assume that the reason most of their geolocators recorded wintering locations in the Bahamas is that most birds *went* to the Bahamas, not realizing that plenty of birds were also heading to Cuba, never to return.*

Then there's the question of how the birds themselves are affected by carrying a device equivalent to 5 percent of their body weight on their backs as they make these impossibly long flights. The data out there is mixed. Overall, the research that has been done on geolocators' effects on small songbirds has found that they do have a small but measurable impact on the survival of the birds that carry them. Unsurprisingly, that impact is greater with bigger geolocators and with smaller birds. Gunnar Kramer was one of the researchers behind a 2018 study on warblers called common yellowthroats that compared geolocator-tagged birds with birds that had only leg bands and found that yellowthroats with geolocators were 60–78 percent less likely to make it back from migration. "Yellowthroats are tiny, and they have short, stubby wings, and for some reason they came back at a lower rate," he told me. There were never any plans to continue studying

---

* The researchers suspect that the lower returns from Cuba are related to the fact that painted buntings are frequently captured in Cuba for the illegal caged bird trade. I talked to the lead author of this study, Clark Rushing, for an article I wrote for the American Bird Conservancy magazine in 2021 on painted bunting conservation, and he described hearing from a contact in Cuba about a couple birds caught by trappers there with "weird little devices" on their backs—his geolocators.

them with geolocators after that initial project, but if there had been, Kramer said that he and his colleagues would have probably reconsidered.

Overall, though, Kramer feels good about the many geolocator studies of warblers he's been involved with. "Being able to understand their ecology and translate that into effective conservation action, that's the most important reason why we would take it upon ourselves to strap something onto these birds," he said. "Yeah, they're carrying 3 to 5 percent of their body mass for a year, and that's a lot. And so we're very thoughtful and careful with how we design these studies and the reasons why we want to collect these data. It really comes down to conservation."

Of course, these questions don't *just* apply to geolocators and warblers, even though warblers are the smallest long-distance migrants to wear tracking devices thus far. Ornithologists are themselves keenly interested in how tracking devices affect birds—not only out of purely ethical considerations (although that's certainly a factor), but also because if the behavior you're trying to study is affected by the device you're using to study it, your results might be meaningless. As a result, there are a *lot* of studies out there related to this topic, so many analyses and meta-analyses that it's hard for one person to get a handle on it all.

It turns out that the oft-cited "5 percent rule," the guideline that a transmitter or data logger's weight should be no more than 5 percent of a bird's total body weight, isn't backed up by any data; it was just a suggestion Bill Cochran made in the early days of the field in the 1960s. Today, many ornithologists aim for a cutoff of 3 percent instead. But when I went hunting for recent reviews of the data on how tracking devices affect birds, two meta-analyses—in other words, analyses of analyses—stood out. One, published in 2019, looked at more than thirty-four hundred studies and concluded that "there was no device mass threshold below which effects were not observed." The other, which came out in 2018, found that as devices have gotten smaller and smaller, the average weight of devices relative to the weight of the birds that carry them has *not* declined. Instead, we've just been putting them on smaller and smaller birds.

I contacted Rob Robinson and Steve Portugal, two of the researchers behind these overviews, to learn more about what all this means. We talked about the murky definition of "effects": although you can theoretically observe some kind of "effect" from even the smallest device, that could be something as subtle and ultimately harmless as a bird engaging in an extra-long preening session while getting used to its new backpack. We talked about the importance of doing studies with control groups, where you compare the survival and behavior of untagged birds with those with various kinds of tags, as Kramer did with the yellowthroats. And we talked about the need to think through the unique vulnerabilities of each bird species and the unique risks posed by each study design instead of relying on a single weight-based rule of thumb.

Eventually, I realized that as thoughtful and helpful as they were, talking to ornithologists wasn't quite enough. I needed to interview a philosopher. A plea on Twitter for an ethicist who could talk me through the considerations involved in putting a backpack on a bird led me to William Lynn, a researcher at Clark University's Marsh Institute for the study of human–environment relations and founder of a think tank dedicated to animal ethics. Lynn led an ethics review to help the U.S. Fish and Wildlife Service decide whether to cull barred owls to protect endangered spotted owls in the Pacific Northwest. He thinks about this stuff for a living.

Not really knowing where to start, I asked him what he would say to an ornithologist who came to him trying to decide whether to go through with a tracking study. He immediately zeroed in on the same question I'd been thinking about: How do we decide when the benefits of the knowledge we gain from these studies outweigh the potential harm to the individual birds?

We can't just say that the individual birds burdened with tracking devices don't matter at all, Lynn insisted. "Those birds are living beings, too. And not just living beings, they're not just like a paramecium, they're self-aware," he said. "They're social, they feel, they think, they make decisions, they have relationships. Those are the reasons we assign ourselves moral value, and that's why birds have moral value, what we call intrinsic value in ethics." (Researchers who study animal cognition may debate whether birds are truly self-aware, but not, I think, whether they have intrinsic value as Lynn describes.)

But an ecosystem has moral value, too, and it's that value that the data collected through these studies can help protect if it's put to good use. It comes down to just thinking through both sides of the equation as fully as you can. "What kind of harms are going to come to these creatures? How many, for how long?" said Lynn. "And then if you put it on a scale, is it going to be justified or unjustified?"

There are no firm answers, only more and more questions to wrestle with. But after spending months immersed in the intricacies of bird tracking tech—reading scientific papers about what we've learned from these devices, browsing the websites of the companies that sell them, talking to the researchers who go into the field to put them on birds—I'm convinced that, by and large, putting tracking devices on birds is worth doing. Yes, there's no way to eliminate the possibility that carrying a transmitter or data logger on its back might make a bird's life a little harder, whether it's a thrush on its way from Colombia to Canada or a warbler making a three-day crossing of the western Atlantic. But every ornithologist I've spoken with cares deeply about the well-being of the individual birds that pass through their hands, expressing a commitment to making sure they're following the latest guidelines for minimizing tracking devices' effects. Any study that involves marking wild birds in any way usually requires multiple levels of review and approval before it can commence as well, requiring permits from the federal government and sign-off from university ethics boards. And crucially, the data collected has the potential to alter the fate of entire species for the better. Someone wise once said that the needs of the many outweigh the needs of the few—or the one.

But while these devices are irreplaceable, some of the latest advances in bird migration research focus on tracing where birds have been by analyzing what ornithologists call "intrinsic markers." Instead of relying on trackers that the birds must carry with them on their travels, today we can follow their journeys using information already encoded in their feathers, their blood, and their DNA.

Seven

You Are Where You Eat

Imagine you're a deuterium atom.

Okay, I realize your eyes might be glazing over just from reading that sentence, but stick with me; we'll be back to birds in a couple paragraphs. You're a deuterium atom, which means you're a very special type of hydrogen atom, one that has an extra particle in your center (or nucleus). And as part of a water molecule, one of the two Hs in H_2O, you, my friend, are part of the world's oceans. You and a few quattuordecillion* of your best friends make up the medium through which narwhals and blobfish and giant squid swim.

Thanks to that extra neutron in your nucleus, you weigh twice as much as the more common variety of hydrogen atom. Only one in about sixty-four hundred hydrogen atoms in the ocean is deuterium. That neutron makes you exceptional.

One day you're churned up to the ocean's surface and you happen

---

* That's a one with forty-five zeroes after it, from a very rough estimate of the total number of water molecules in the oceans.

to evaporate. Now you're in a cloud (whee!). But with that neutron weighing you down, it's hard to stay airborne. Shortly after your cloud moves over land, you find yourself falling back to the surface as rain, even as some of your lighter companions sail onward. You seep into swampy soil and then enter the roots of a bald cypress tree, which eventually incorporates you into one of the lacy needles that line its branches. A passing caterpillar munches you up. And then a bird eats the caterpillar.

See, I told you we'd get back to birds.

The bird is a male prothonotary warbler, a "swamp canary" whose cheerful yellow plumage contrasts with its blue-gray wings and tail. This bird is currently undergoing its annual molt, growing an entirely new set of feathers and shedding its old ones in preparation for its migration southward. Once you were tumbled in the eddies created by a passing manta ray; now you and your neutron find yourselves lodged in a freshly grown "primary" feather, one of the long flight feathers along the edge of the wing, ready to depart for South America with the feather's owner.

And your journey is about to take one more unexpected twist. When the prothonotary warbler stops for a rest in a patch of Panamanian rain forest, he blunders into a mist net. A researcher gently plucks the feather you're a part of and tucks it, and you, into an envelope. You're on your way to a lab, where you'll become part of an effort to learn more about this warbler species' migratory habits and maybe even improve its chances at survival, via a scientific technique called stable isotope analysis—a method for studying migration that's less invasive than putting a backpack on a tiny bird and lets scientists gather data on hundreds or thousands of birds inexpensively and efficiently.

Mapping the Isoscape

One of the first people to realize that isotopes might help reveal where migrating birds had begun their journeys was Keith Hobson. Hobson, a research scientist at Environment Canada, has an undergraduate degree in what might seem an unlikely field for someone who today is known primarily for his work on bird migration: physics.

"It's a curious path, which I recommend nobody does," he said drily when I got him on the phone. "I'd spend my weekends bird-watching, but [physics] was my discipline."

Physics led him to isotopes, which are essentially different varieties of a single element from the periodic table. All atoms of a certain element have the same number of protons in their nuclei. Hydrogen has one proton, which is what gives it the number one slot in the periodic table. The different isotopes of hydrogen *all* have just a single proton at their cores, which is what makes them hydrogen. But there's a second particle that can be found in the nuclei of atoms, called a neutron, and the number of neutrons is where isotopes differ. The vast majority of hydrogen atoms don't have any neutrons at all, but a few—the ones we call deuterium—have one. An even smaller fraction, which we call tritium, have two.

After finishing his physics degree, "I worked for a number of years in a radiocarbon lab," said Hobson, "dating artifacts and things." Radiocarbon dating, which you might have heard of in connection with archaeology, is based on analyzing the amount of a radioactive isotope of carbon, carbon-14 (14 for its combination of 6 protons and 8 neutrons), in a sample. Being radioactive means that carbon-14 "decays" over time, shedding particles and transforming into nitrogen. (This is why radioactivity is sometimes dangerous: those shed particles can wreak havoc if they pass through organic tissue.) When a plant or animal is alive, the amount of carbon-14 in its tissues is constantly refreshed from the environment as the organism takes in air or food. Once a plant or animal dies, however, whatever is there begins to decay at a constant rate. We can tell roughly how long this process has been going on by measuring how much carbon-14 is left.

But while radioactive isotopes like carbon-14 can be used to determine an object's origins in *time*, stable isotopes can often help determine where an object originated in *space*. "Stable" means that these isotopes aren't radioactive; they stay the same over time instead of shedding particles and turning into something else. Carbon, for example, has multiple stable, nonradioactive isotopes in addition to unstable carbon-14. The relative amount of each of those two stable carbon isotopes in the plant life of a specific place depends on how dry the local environment is. Plants in dry habitats have evolved a more water-efficient way of doing photosynthesis, and this happens to result in slightly more of the rare stable isotope carbon-13, which also shows up in the tissues of people and animals that consume those plants. Relationships like this have helped scientists determine everything from the movements of ancient human populations to the geographic origins of poached elephant ivory.

"Way back then in the dark ages, the archaeologists were ahead of everyone in this field, and it just dawned on me, why aren't biologists using these techniques?" said Hobson. When he eventually returned to school for a doctoral degree, he decided it was time to bring his two passions together.

Hobson's PhD, completed in 1992, focused on food webs in the high Arctic. Unusual stable isotopes of carbon and nitrogen become slightly more concentrated as they work their way up the food chain, and by analyzing samples of muscle tissue, Hobson was able to reconstruct the predator–prey relationships among plankton, fish, seabirds, and even narwhals and polar bears.

But while carbon and nitrogen were useful for figuring out what birds had been eating, it was hydrogen that finally showed the power of isotopes for figuring out where they were going.

Hydrologists had known since the 1960s that the amount of deuterium in the environment varied across the North American landscape. The farther north you go, from the Gulf of Mexico toward Canada, the tinier the amount of deuterium in the local water becomes. The reason, as Hobson explained to me, is the way deuterium is weighed down by its bonus neutron. As moisture sweeps inland from the Gulf, the heaviest water molecules—the ones containing deuterium—tend to fall as rain first. The farther north you get, the less deuterium is left in the rain, creating what scientists call a "latitudinal gradient."

Hobson joined Environment Canada shortly after finishing his PhD, and the deuterium gradient captured his interest almost immediately. "I conferred with my colleague," he told me, "and I said, do you think these patterns could be transferred to organisms?" In other words, if a bird spends time in a given environment, eating and drinking and therefore ingesting water with a certain amount of deuterium, would the specific ratio of normal hydrogen to deuterium for that location then be reflected in the bird's body itself? "People were very skeptical," said Hobson, "that a pattern in rainfall averaged over an entire year or growing season could manifest itself continentally in what we're now calling isoscapes" (a mash-up of the words "isotope" and "landscape").

He and his hydrologist colleague Len Wassenaar set about proving it was possible. First, they plotted the average amount of deuterium in the rain at different locations across North America, producing something resembling a contour map. Then they spent two years collecting feathers from six migratory songbird species at fourteen different sites across the continent, from Swainson's thrushes in Alaska to American redstarts in Louisiana. (Removing a single feather doesn't noticeably impede a bird's ability to fly. Hobson and Wassenaar didn't do the field work to collect all those feathers themselves, but instead recruited other researchers who were willing to mail them samples from various locations. "There was a lot of networking," recalled Hobson.) Analyzing the hydrogen contents of those feathers—more than 170 in all—and placing them on the map showed that yes, the amount of deuterium in a feather reflected the average amount of deuterium in the environment at the location where it was grown. "Bingo. There it was staring at us," said Hobson.

So far, so good. But those feathers came from birds captured on their breeding grounds. Could isotope analysis tell a scientist where a bird had *been*, even when it wasn't there anymore? Migratory songbirds in the Americas typically do their annual feather molt on their breeding grounds, growing a new set of feathers there that they'll keep throughout their migration and the winter season. And unlike the hydrogen in, say, blood or muscle, which is regularly replenished from the local environment as the bird is on the move, the isotopes in feathers are "locked in" once they're grown. That means that the feathers of a bird on migration should still carry the isotopic signature of the place where they were grown.

To test this, Hobson and Wassenaar obtained feathers from migrants that had spent the winter of 1993 in Guatemala. They chose three species—the worm-eating warbler, hooded warbler, and wood thrush—that breed mostly in the southeastern United States and two—the ovenbird and gray catbird—that also range into Canada. If their isotope map placed the origin of a hooded warbler feather in, say, Ontario, they'd know something wasn't working. For every feather they checked, however, the feather hydrogen isotopes placed the bird at a point on the map of North America consistent with the species' actual breeding range. The method worked.

At the same time, a second research team headed by the Stanford University geophysicist Page Chamberlain was carrying out a parallel study focused on a single species, the black-throated blue warbler. Like Hobson's work, that project confirmed that the amount of deuterium in the warblers' feathers varied across their range and that this could be used to estimate where warblers wintering in the Caribbean had spent the summer. After some initial skepticism from scientific journal editors, both studies were published in 1996. "And then you could say the rest is history," said Hobson. "People went crazy applying this technique."

Sometimes on their own, sometimes in combination with transmitters or light-level geolocators, the methods Hobson pioneered have been used to study the movements of birds ranging from the familiar American crow to the mysterious Vaux's swift, which flies continuously when not nesting or roosting. But even after spending hours talking to Hobson, I still had questions about how these techniques work in practice—how an ornithologist gets from a plucked feather to a number representing the feather's deuterium content.

It was time to head to a lab.

Above: Samuel Prentiss Baldwin prepares to band a bird captured in one of his traps, ca. 1919.

Bottom left: The "Pfeilstorch" shot near Rostock, Germany, in 1822.
Bottom right: Leon Cole in his laboratory, date unknown.

Bird banders at Powdermill Nature Reserve outside Rector, Pennsylvania, measure a Wilson's warbler.

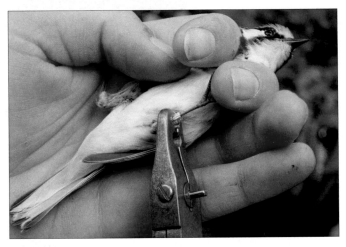

A chestnut-sided warbler is banded at Powdermill Nature Reserve.

Annie Lindsay bands a bird at Powdermill Nature Reserve.

Birds captured at Powdermill Nature Reserve are kept in cloth bags as they await their turn to be banded.

Bird bands must come in sizes to fit all birds, from hummingbirds to eagles.

Bottom left: A just-banded northern flicker is safely immobilized in a tube so it can be weighed before being released at Powdermill Nature Reserve.
Bottom right: A cedar waxwing about to be extracted from a mist net at Powdermill Nature Reserve.

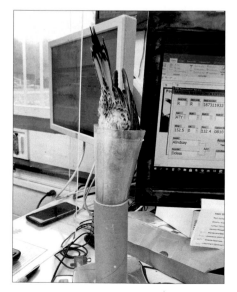

Richard Graber stands next to the parabolic reflector and recorder that he and Bill Cochran used to record full nights of flight calls in central Illinois in 1957 and 1958.

Above: Spotting scopes outfitted with the LunAero system for automated moon-watching.
Left: Martin Minařík poses with the recording equipment he uses for his nocturnal flight call hobby.

Above: The author tries out moon-watching in her driveway in Walla Walla, Washington, during spring migration in April 2021.
Right: One of the autonomous recording units for long-term monitoring of nocturnal flight calls during migration at MPG Ranch outside Florence, Montana.

Debbie Leick checks on the recording equipment at one of the autonomous recording units at MPG Ranch in August 2021.

One of the recording stations in MPG Ranch's Bitterroot Array perches atop the Willow Mountain fire lookout in Montana's Bitterroot National Forest in January 2019.

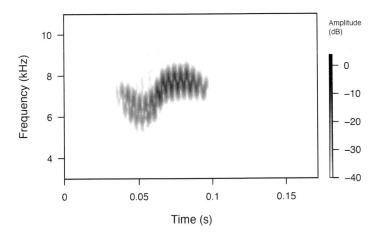

A sound spectrogram of the nocturnal flight call of a black-and-white warbler.

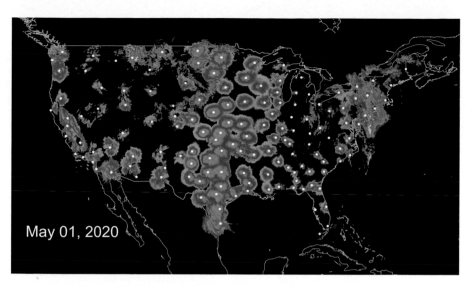

May 01, 2020

Large numbers of migrating birds appear as blue blobs on North American weather radar.

Sidney Gauthreaux processing radar images at Louisiana's Lake Charles Weather Station in May 1996.

The "porcupine" plane that Richard Graber used to track radio-tagged thrushes in the 1960s, with pilot Jim Taylor in the cockpit.

One of Ana Gonzalez's Motus towers in the Colombian Andes.

A Swainson's thrush being fitted with a harness and a Motus transmitter in March 2015 for Ana Gonzalez's research in Colombia.

Left: Auriel Fournier with one of the Motus towers she uses to monitor rail movements near Havana, Illinois, in September 2021.
Right: A sora tagged with a Motus transmitter, ready to be released for Auriel Fournier's research.

Auriel Fournier and her colleagues work to attach a Motus tag to a sora near Havana, Illinois, in September 2021.

A prototype of the Terra device, which will allow anyone to pick up Motus-tagged birds flying over their yard.

Left: A bar-tailed godwit with a satellite transmitter antenna visible after arriving in Alaska. *Right:* The first spoon-billed sandpiper tagged with a satellite transmitter in October 2016.

Left: Andy Boyce measures the bill of a long-billed curlew held by Paula Cimprich near Zortman, Montana, in May 2021. *Right:* Andy Boyce and intern Mary De Aquino prepare to release a curlew just fitted with a GPS device.

Lucy Hawkes and Nyambayar Batbayar pose with a bar-headed goose tagged with a GPS transmitter in Mongolia in July 2011.

Field assistant Sara Douglas attaches a light-level geolocator to a Connecticut warbler east of Winnipeg, Manitoba, in May 2016, part of Emily McKinnon's research on their movements.

Left: A Connecticut warbler wearing a light-level geolocator.
Bottom left: A "furious" Magellanic penguin with one of Rory Wilson's early light-level geolocation devices on its lower back in 2014.
Bottom right: A geolocator-carrying a blue-winged warbler from Gunnar Kramer's research, just returned to its breeding site in Kentucky, May 2016.

Left: An array of geolocators with leg-loop harnesses, laid out in a hotel conference room in Tennessee, as Gunnar Kramer and his research assistants prepared for field work in April 2015.

Right: Garrett Rhyne works to gently remove a geolocator from a recaptured Swainson's warbler north of Baton Rouge, Louisiana, in April 2022.

A happy Garrett Rhyne shows off his recaptured Swainson's warbler.

Grad student Sadie Ranck demonstrates the process of preparing a feather or claw sample for stable isotope analysis on the campus of Idaho's Boise State University in November 2021.

A feather sample stuffed into a tiny silver capsule, ready to be run through a mass spectrometer.

The series of machines that break a feather or claw sample down into its constituent elements and measure their relative amounts for stable isotope analysis.

Boise State University professor Linda Reynard checks the computer readout from the mass spectrometer, showing the amounts of deuterium and "normal" hydrogen in a feather sample.

Martin nests (the dark clumps in the shadows) under the eaves of Theunis Piersma's house in the Dutch village of Gaast in May 2021.

Boise State University researchers gently clip a sample from the claw of an American kestrel in June 2015.

An artist's rendering of the blackpoll warbler museum specimens used in Camila Gómez's stable isotope study.

The "freezer farm" in a basement on the campus of Colorado State University in Fort Collins holds tens of thousands of the Bird Genoscape Project's feather and tissue samples.

American Robin (or "AMRO") feathers in the Bird Genoscape Project's freezer storage.

Chandler Robbins on a Breeding Bird Survey in the 1980s.

Avid eBirder Bruce Toews scans for winter birds outside Walla Walla, Washington, in February 2022.

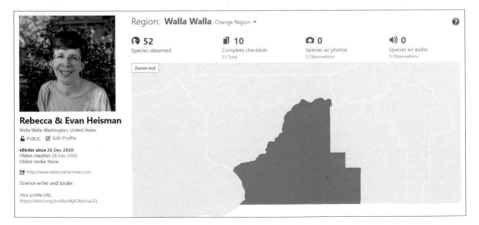

A screenshot of the author's eBird profile.

Analyzing a Feather

As I drove into Boise, Idaho, on a cool November morning, a beautiful sunrise spread over the mountains in front of me. Instead of a nature preserve or national forest, however, my destination was a pair of small rooms off a hallway in Boise State University's Environmental Research Building. I'd seen scientists pluck feathers and tuck them into envelopes while handling birds in the field, but I couldn't really picture what happened after those feathers were whisked off to a lab like Hobson's. Without seeing that end of the process, I felt as if I were only telling half the story of how stable isotope analysis works. Luckily, a pair of Boise State scientists had agreed to show me where those feathers go next.

Linda Reynard, a faculty member in the university's geoscience department and head of the lab, and Sadie Ranck, a master's degree student studying American kestrel migration, met me in the building's lobby. Due to the COVID pandemic, all three of us were wearing face masks, but Ranck's was particularly striking, the fabric printed with colorful blue jays and nuthatches. When I told her I liked it, she admitted that her mom had sewn her a whole series of bird-patterned masks in honor of her research.

Ranck started working in the BSU ornithologist Julie Heath's lab as an undergraduate, monitoring kestrel nest boxes, and stayed on to study the migration patterns of the small falcons as a graduate student. Kestrels in Idaho's Treasure Valley are "partially" migratory, meaning some head south for the winter while others stay put. Ornithologists have come up with multiple hypotheses to explain partial migration; maybe bigger birds stick around because they can tolerate the cold better, or maybe males stay so that they can claim and defend good territories before females come back in the spring. But prior to Ranck's project, researchers didn't know a lot about which kestrels in the Boise area stayed, or why, or whether individuals stuck with the same strategy throughout their lives or could switch. "We don't know what's driving those patterns—who's deciding to stay, who's deciding to go," said Ranck.

The schedule on which kestrels molt and regrow their feathers is less straightforward than it is for songbirds, so instead of feathers Ranck uses claw samples to determine whether an individual bird is resident or migratory. The birds' claws regenerate every four to six months, "so we clip a millimeter or two off the very tip of their outer claw on the right foot, and that signature reliably reflects the wintering area. It's pretty straightforward, actually." It helps that Ranck doesn't need to know *where* exactly the migratory kestrels in her study went. She just needs to know whether they migrated or not, and using the techniques developed by Hobson, she can look at whether or not the amount of deuterium in a kestrel's claw accumulated over the past few months matches up with the local Boise environment.

On the day I visited, Ranck wasn't actually preparing any claw samples, but she walked me through the process, beginning with carefully washing them with a series of chemicals. Another student working with barn owl feathers had left behind some spares, and Ranck and Reynard let me handle the tiny plastic vials that contained the feather clippings. The eventual destination of samples that enter the lab is a series of machines in the room next door that break them down and analyze their contents, but we had another step to complete first.

On a table in the small back hallway connecting the two sides of the lab sat a microbalance—a scale designed to weigh such minute quantities that you need to insert whatever you're trying to weigh through a little trapdoor that you then close, lest air currents in the room throw off the delicate precision required. Ranck showed me a pill-bottle-like container full of silver cylinders each smaller than my little fingernail, then demonstrated the process of slicing off an almost invisibly tiny wisp of feather with a razor blade, jamming it into the silver capsule with a pair of delicate tweezers, crumpling the whole thing into a little ball perhaps the size of a sunflower seed, and then placing it on the microbalance to make sure it contains the right amount of material, between 0.3 and 0.4 milligrams. That's 0.000014 ounces, so small a weight I'm having a hard time finding anything to compare it to. Have you ever had a single teeny tiny down feather come out of a seam in a coat? It's about a third of the weight of one of those.

Each sample needs to be encapsulated like this to keep it together in a neat package as it passes through the machinery, but it is very, very fiddly work. Ranck let me sit down and give it a try myself, inserting a second feather bit into another silver cylinder. At first I thought I'd managed it, jabbing awkwardly with the tweezers, but when I put my own tiny foil ball on the microbalance it turned out there was nothing in it. I'd somehow dropped my feather fragment while trying to maneuver it into the capsule.

Ranck laughed sympathetically. Her claw samples, she explained, could be even trickier. Being more "poky" than feathers, as she put it, they would sometimes puncture the little silver capsule and she'd need to start over. "There's actually a whole website with pictures of how not to fold the ball," she said. "Like, don't fold it into a pyramid, don't fold it into a potato chip." The memory of preparing these silver balls seems to haunt every ornithologist who's ever worked with stable isotopes; when I posted a photo of the process to Twitter after my trip, I received a range of replies that all essentially boiled down to "UGH." Ranck told me she could get through about fifty capsules in a day and prepared a total of 880 of them for her thesis research.

Once they're ready, the capsules get loaded into individual slots in the carousel of a piece of equipment called a thermal conversion element analyzer. One by one, they drop down into the machine and are "pyrolized," subjected to such high heat that they vaporize into a collection of hydrogen, carbon monoxide, and other trace gases. (Reynard explained that this is different from simply incinerating them, because combustion requires the addition of oxygen, which could throw off the isotope analysis.) These gases then go through another machine called a gas chromatograph, which separates them and sends the hydrogen and carbon monoxide from each sample into the mass spectrometer.

Each of these three instruments is just a large gray box, the sort of thing that at a casual glance might be mistaken for, say, a photocopier. But inside the mass spectrometer, a magnetic field sorts the particles of each gas by weight. Lighter particles feel the tug of the magnetic field more strongly than heavier ones, so that deuterium hits one detector and "normal" hydrogen hits another. The measurements from these detectors appear as a series of graphs on a computer screen, and there it is: the ratio that provides a vital clue about where each bird has been. Each sample takes about ten hours to run, from the time it enters the first machine to the time the mass spectrometer spits out the results, but because they drop from the carousel into the first piece of equipment at regular intervals, a new set of numbers appears around every twelve minutes.

When she pored over these numbers, Ranck's long days spent preparing samples paid off. "We found that males are more likely to migrate than females, which supports the body size hypothesis, because females are around 9 percent larger than males," said Ranck. "And we also found that in colder-than-average winters, smaller birds are more likely to migrate than larger birds." So kestrels can indeed switch strategies in response to what's going on in their environment.

But winters in the Treasure Valley are getting warmer, and as a result fewer and fewer kestrels are choosing to migrate. Whether there will be enough prey to go around if the entire population stays put over the winter is yet to be seen. Further complicating things, raptor biologists often use the number of birds counted overhead during migration as a snapshot of how a species is faring from one year to the next, and a change in the percentage of birds that migrate could make it hard to interpret that data. Kestrel populations are in decline across much of North America, and having an accurate idea of how they're faring over time is crucial for conservation efforts.

When I mentioned to Ranck that I'd spoken to Keith Hobson, her eyes lit up; although she'd never met him, she'd cited his work again and again in her own studies. Hobson has had a hand in more ornithology projects than he can probably remember. But when I'd asked him if he had any particular favorites, the first one he mentioned was not about birds but about insects.

"Solving a Migration Riddle"

Like many bird-watchers, Hobson has a secondary interest in butterflies, and one of the first species he and Len Wassenaar applied their new deuterium analysis to was the famously migratory monarch.

The monarch butterflies of eastern North America migrate to a small patch of Mexico's Sierra Madre for the winter. (A second monarch population in western North America winters in California.) Butterflies, of course, have much shorter life spans than birds, and this annual migration spans multiple generations; along the way, monarchs stop to reproduce, laying eggs on milkweed plants that will eventually carry on the journey as adult butterflies. Researchers had attempted to track the movements of individual monarchs using tiny adhesive tags on their wings, but observations of tagged butterflies after they were released were scarce.

By analyzing the carbon and hydrogen isotopes in the tissues of monarchs wintering in Mexico, however, Hobson and Wassenaar were able to determine in the late 1990s that most of them had started their lives in the American Midwest—the "corn belt." According to Hobson, this surprised some butterfly scientists, who didn't think there was enough of the milkweed the butterflies required in that region. "People were like, what are you talking about? These things need milkweed plants!" said Hobson. But historically, there *had* been plenty of milkweed in the farm fields of the Midwest, growing between and underneath the swaying rows of corn.

In 1998—the same year Hobson's monarch paper was published— the company Monsanto introduced corn genetically modified to tolerate the spraying of harsh herbicides that eliminated all other plants in a field, including milkweed. Since then, the eastern monarch population has crashed, falling by more than 80 percent. Hobson's work was an early alarm about the importance of the Midwest for this species, and conservation groups now encourage people in the region to plant milkweed wherever they can to help boost the butterflies' chances at reproducing successfully.

But Hobson does have a favorite bird project as well.

Many birds migrate between Europe and Africa, a parallel to the species that travel between North America and the Neotropics. For some reason, however, Europe-to-Africa songbird migrants typically molt their feathers during the winter, growing a new set in Africa before returning to Europe, the opposite of what North American migrants do. This means that in theory you could take a feather sample from a bird in Europe, analyze the amounts of various isotopes within, and figure out roughly where the bird spent the winter. Since on-the-ground field work can be difficult in many parts of Africa, this would be great information for scientists studying those species to have.

Hobson had been corresponding with many ornithologists in Europe experimenting with stable isotopes. The problem, in part, was that how deuterium varies across the African landscape wasn't fully clear, "simply because you need many, many years of collecting water and having it analyzed," he told me. What data there was showed that because of differences in weather patterns there was no tidy gradient like in North America.

But hydrogen, of course, isn't the only element in the environment with multiple stable isotopes. Hobson and other scientists had also been using carbon, nitrogen, and others for years. "What really hit me," Hobson said, "was the fact that plant physiologists had already tried to map what plants would look like isotopically across Africa." Remember, the amount of carbon-13 in plant tissue depends on what chemical reactions the plant relies on for photosynthesis—something heavily influenced by how dry the local environment is. The amount of nitrogen-15 in plants also varies with the local climate. "They have a map of Africa, they have a map of the whole world, and what they think [the isotope ratios in] plants should look like. So I looked at those maps and I thought, well, wait a minute, why don't we use these maps then to predict what a feather would look like if it was grown at those locations?"

In 2012, Hobson, Wassenaar, and their colleagues published a study presenting a "multi-isotope feather isoscape" for all of Africa, combining what was known about deuterium in Africa at the time with plant physiologists' maps of carbon-13 and nitrogen-15. By analyzing the amounts of all three isotopes in a feather and looking at where on the continent those three specific isotope concentrations occurred in combination, ornithologists could now narrow down the bird's wintering location considerably.

It wasn't perfect. Sometimes a specific combination of concentrations of the three isotopes Hobson focused on wasn't unique, occurring at two or three different spots around the continent. Ornithologists would then need to use other information about the bird, like scattered reports from bird-watchers or data on what sort of habitat it preferred, to make an educated guess about which was its true wintering location. And surveying birds on the ground in Africa to prove that the results were correct was likely to be nearly impossible for many species. "People could say, well, what if you're wrong?" said Hobson. "And my point would be, yeah, but we know a heck of a lot more now than we did before without this technique. And this is now a part of the toolbox."

That tool was just what Theunis Piersma needed to solve a mystery that was, literally, right outside his door.

Piersma, a Dutch ornithologist, is one of the world's leading experts in shorebird migration. After meeting at an ornithological conference, he and Hobson had kept in touch over the years, corresponding about their shared interest in the movements of birds. But in addition to his day job tracking godwits and red knots up and down the coastlines of the world, Piersma had developed an ornithological hobby: studying the martins that nested in the eaves of his house in a rural part of the northern Netherlands. I felt a little awkward contacting Piersma and telling him that I wanted to interview him about his martins rather than the migrating shorebird work he's most known for, but he was eager to talk about them.

Piersma bought his home, a former primary school built in 1956, in 2001. "One of its attractions was that it had a big house martin colony under one of the eaves," he told me. "But of course, the guy that was selling the house didn't realize that, so he carefully removed all the house martin nests before he put it on sale, which meant we had to start from scratch."

It didn't take long for the birds to return. House martins, relatives of North American swallows, build colonies of mud nests under bridges and on cliffs as well as on the eaves of buildings. Martins and swallows belong to a loose category of birds known as aerial insectivores, species that catch insects on the wing, which have been declining worldwide in recent decades, possibly due to widespread pesticide use reducing the numbers of the bugs they depend on. There are still so many martins in Europe that most people continue to think of them as common, but Piersma estimates that their populations might have declined by 98 percent since World War II.

Piersma's new home was in a small village called Gaast, and his new neighbors quickly noticed his interest in the martins that fluttered between their houses in summer. "There was this one lady who was really interested," he said. "I met her on the street the third year we were living here, and she told me, the house martins are back, I saw the first one yesterday, that's two days earlier than last year." It turned out she had been recording the date that the first martin reappeared in the town each spring for years. Knowing that she was talking to an ornithologist, she started quizzing him, asking him where the birds spent the winter. "And honestly, we didn't know at all at that point. [They go] into Africa and it's kind of a riddle, where they disappear to."

Few banded house martins are recovered on their wintering grounds in Africa. But, Piersma realized, Keith Hobson's new African isoscape offered a possible way to answer his neighbor's question. All he would need were some feathers.

Piersma isn't really a fan of capturing and handling birds, even though he acknowledges it's often necessary; he dislikes the feeling that he's causing the animal stress. Still, he admits that his house martin "field work," having breakfast and then strolling out onto his porch to catch some birds, was pretty nice. Sending the feathers off to Hobson for isotope analysis, he awaited the results.

The resulting paper, with a title that began "Solving a Migration Riddle," was published in 2012. It gave Piersma his first-ever peek at where "his" martins were going when they disappeared from his porch in the fall: they migrated three thousand miles to the west coast of Africa, particularly Cameroon and the Congo basin, going from delighting Dutch villagers to sharing habitat with gorillas and forest elephants. Piersma notes that the deuterium part of the analysis didn't end up being very useful, due to some effect related to El Niño precipitation the year of their study. "That appeals to me, actually, that here in the Netherlands, at your house in a little village, you can measure El Niño effects in Africa!"

But Piersma's neighbor, the one whose question he'd set out to answer, didn't subscribe to scholarly journals. So Piersma ultimately wrote a book about his house martin riddle—not for his fellow scientists, but for his fellow villagers in Gaast, where the locals speak not Dutch but the minority language Frisian. Eventually, it was translated into Dutch and English, but the original Frisian version of *Guests of Summer: A House Martin Love Story* sold around six thousand copies, making it a bestseller by the standards of a language in which few books are written.

"For that little book, I've had maybe ten times more response than for all my other work," Piersma said happily.

Bottlenecks and Time Travel

As Piersma found, part of the power of stable isotope research comes from the fact that—compared with the intensive work necessary to track songbirds with geolocators—it is both easy and inexpensive. No need to buy high-tech devices and attach them to the birds; no need to recapture the birds the following year. As a result, it's given scientists new insights into the flow of bird migration through regions that can be tough to work in.

Along the border between Panama and Colombia—which is also the border between Central and South America—lies what is known as the Darién Gap. Consisting of marshes and mountainous rain forests, this region gets the "gap" part of its name from the fact that there are no roads that traverse it. The Pan-American Highway, a network of roadways that otherwise extends from Alaska to Argentina, simply stops existing for sixty-six miles when it gets to the Darién and then picks up on the other side. You can't get there from here.

In 2010, the ornithologist Nick Bayly set up a project to study the birds that pass through this region during migration. "Picking that area to do a study was a bit of a no-brainer, because anybody who looks at a map is going to see that there should be a natural bottleneck," Bayly told me via video call from his home in Colombia. It makes sense that birds following an overland migratory route to get to South America would be funneled through the Darién in large numbers. "But it had never been studied until that point," largely because of the region's "wild west" reputation. Not only was the physical environment challenging, but the Darién Gap was (and is) also infamous as a hot spot for the smuggling of both migrants and illicit goods.

After completing his PhD, Bayly had relocated from the U.K. to Colombia in 2008, where he became one of the founders of the Neo-tropical conservation nonprofit SELVA. When he first arranged to do field work at a private nature reserve on the Darién's Panamanian side, he and his colleagues weren't thinking about doing stable isotope analysis; they were just counting every bird that passed through during migration. They ran mist nets every day for two months. They walked transects through the rain forest, counting every bird they saw or heard. And three times a day, they simply stood still for an hour or more and attempted to count and identify every bird passing overhead—and there were a *lot* of birds. "We've managed to count 1.5 million raptors in one season," said Bayly. It was "just birds, all day, every day, as much as you could cope with. You ended up dreaming of little black dots flying across your pupils."

Bayly and his colleagues weren't camping in the rain forest; their home base was a fishing village on the Panamanian coast that was beginning to attract increasing numbers of tourists. "But all the same, the conditions are tough," said Bayly. The climate is extremely humid, and he described sweating from the moment he got out of bed in the morning, sweat dripping down his face as he pulled birds from mist nets and fitted them with bands. "It's difficult to keep going under those conditions, but we could be banding one hundred, two hundred birds in the mornings, and you had to be working flat out the whole morning to make sure you were processing all the birds safely."

They'd been collecting feathers since day one out of thoroughness, but as time went on, Bayly and his colleagues became increasingly curious about where in North America all these birds were *coming* from. They knew there were hundreds or thousands of thrushes and warblers passing through, but they didn't know whether those birds had started out from across the full breadth of their species' breeding ranges or only from small areas: Was the Darién Gap a bottleneck through which the nearly total global populations of entire species were passing every year?

Laura Cárdenas-Ortiz, a Colombian graduate student working with Keith Hobson, analyzed the hydrogen isotopes in more than fifteen hundred feather samples collected in the Darién Gap from eleven songbird species over five years to find out roughly where within their North American breeding ranges they'd originated. The results were dramatic. For six of the eleven species, individuals passing through the Darién Gap were from locations spanning at least three-quarters of their species' total breeding ranges. For the prothonotary warbler, the species that housed our hypothetical deuterium atom in the opening section of this chapter, that proportion approached 100 percent. Almost every prothonotary warbler in the world may pass through this narrow strip of jungle in the fall on its way from the western forests of the southeastern United States to the mangrove swamps on the northern fringes of South America.

This means, of course, that adverse events in the Darién—whether natural, like hurricanes, or human caused, like the logging of the rain forest—could have an outsized impact on birds that nest across large swaths of North America, dimming the beloved spectacle of migratory songbirds' spring return. Bayly hopes that studies like this one will "amplify the call that that region needs more protection and more investment, from the Colombian state and the Panamanian state but also hopefully from actors outside who may be able to use their influence to change the status quo in our region a little bit."

When Bayly started his work in the Darién Gap in 2010, no one had yet put a light-level geolocator on a warbler. By the time the stable isotope study of the bottleneck came out in 2020, however, geolocator studies had been published on at least half a dozen wood warbler species. Before we signed off, I asked Bayly why he thinks stable isotope analysis is still a valuable tool for ornithologists studying migration.

"The way I see it, it's all about scale," he said. Although some studies have attempted to put geolocators or other tracking devices on birds from multiple populations across a species, such projects are incredibly expensive, and therefore few in number. With stable isotopes, you can fairly easily sample hundreds if not thousands of birds in multiple locations and start to piece together that connectivity at a fraction of the cost. Not to mention, of course, that plucking a single feather can be much less invasive than expecting a tiny bird to carry a backpack from one continent to another. "These new technologies are so much sexier, the results are so much more visually impacting, but I personally don't think we should turn our back on stable isotopes. They still have an awful lot to tell us."

They can even help us travel through time. You can't climb into a time machine and put a geolocator on a warbler from the 1970s, but you *can* analyze the hydrogen isotopes in the feathers of a bird that lived decades ago.

Like Nick Bayly, Camila Gómez is another founding member of SELVA. (*Selva* is a Spanish word for "jungle," but according to Bayly it has more "spiritual" connotations than the English word.) Gómez has loved birds ever since she started tagging along on bird-watching outings with her grandfather as a young girl in Colombia. As a PhD student at Colombia's Universidad de los Andes, she collaborated with Keith Hobson and was introduced to stable isotopes, but the idea to use them as a window into the past originated through her work on the Colombia Resurvey Project. This initiative, which aims to re-visit sites in Colombia originally surveyed by the famed ornithologist Frank Chapman in the 1910s and document how the bird communities have changed, started her thinking about other ways to compare birds across time. Having worked with Hobson, she quickly realized that stable isotopes might offer another avenue to do this.

Natural history museums—the same places we take our kids to see dinosaur skeletons—often house vast collections of animals killed and preserved in the name of science. In back rooms out of the public eye are jars of fish, trays of pinned insects, drawer after drawer full of colorful birds. Although the days when a shotgun was a key part of every ornithologist's tool kit are long gone (and rightly so), scientists do continue to judiciously collect such "specimens," providing a crucial archive of genetic and ecological information. Natural history collections often end up providing a trove of material to study subjects that weren't even conceived of when the animals within them were alive, such as the long-term effects of pollution and climate change.

But the potential of such collections for stable isotope research, still with the deuterium signatures of past molting locations locked into specimens' feathers, remained largely unexplored. The only previous attempt to use stable isotope analysis on bird specimens in natural history collections, a study of rusty blackbirds published in 2010, had found little difference in the deuterium content of historical and contemporary feathers.

"When we spoke initially to curators [at big museums] like the American Museum of Natural History about whether they would give us feathers from their historical specimens, they were like, welllllll, these specimens are really valuable. You can't just pluck a feather from a hundred-year-old specimen, you have to justify this," Gómez told me. "They're a bit cautious, and I understand completely, because those specimens *are* super valuable. So when we found this box of specimens that were readily available and we could use, we said, well, let's try it out."

The specimens in question were a set of preserved blackpoll warblers in the museum of the Universidad de los Andes, where her former PhD advisor, Daniel Cadena, was a professor. They were collected between 1972 and 1975 by Cornelis Johannes Marinkelle, a Dutch zoologist who joined the faculty of the Universidad de los Andes in 1963. A medical doctor as well as a biologist, he was an especially avid collector of eggs, adding more than twenty-five thousand eggs from two thousand bird species to the university's collection during his career. Nine species and subspecies, including a frog, a bat, and a mosquito, have been named in his honor.

Annual bird surveys in North America suggest that the total number of blackpoll warblers, our ocean-crossing friends from the last chapter, has been dropping by about 4 percent per year since the 1960s—a staggering rate over so many decades. Although a single culprit is hard to pin down, the boreal forest in which these birds breed has also gradually been retreating as the climate changes, and scientists forecast that even a few degrees of warming could eliminate vast swaths of their preferred habitat.

At the university museum, Gómez opened the drawer that contained the decades-old blackpoll warbler specimens and delicately removed a single feather from each one. (When I pressed her for details about what it was like to work with the historical specimens—did they have any sort of odor? did she need to wear gloves?—she said it was actually "not very exciting." They were stored in an insulated cupboard and had no smell that she could recall, and gloves are actually frowned upon because feathers can stick to them. "Plucking the feather is no big deal, you just pull firmly, making sure you hold the wing steady.") Meanwhile, Nick Bayly and others captured and collected feathers from fifty-one blackpoll warblers at the same location in the eastern foothills of the Andes where Marinkelle had worked forty-five years earlier. All of the samples were sent off to Keith Hobson's lab in Canada for analysis.

When the ratios of deuterium to normal hydrogen in feathers were plotted on the isoscape map of North America, Gómez was amazed. "This was super different," she said. There was a clear difference between the blackpoll warblers of 1975 and the blackpoll warblers of 2019. In that time, the birds' breeding range had shifted north by more than 370 miles, pushing from the southern tip of Hudson Bay far up its eastern and western shores. "It's probably a mixture of climate change and just human transformation of the southern edge of the boreal forest" through activities such as logging, said Gómez. "That has impacted populations of blackpoll warblers, and it's pushing them north."

This study may only be the beginning. Thanks to the Colombia Resurvey Project, scientists now have lots of new specimens from different areas in Colombia to compare with historical specimens. Gómez and her colleagues hope to do a wider comparison of the Chapman specimens collected 110 years ago and feathers collected recently. In addition to stable isotopes, they plan to look at mercury concentrations in historical versus modern feather samples, another changing factor in the environment that could be affecting the survival of migratory birds.

As I write this, it's mid-October, still prime fall migration time for songbirds. At this moment, some blackpoll warblers are preparing to depart from North America's east coast, others are over open water, and still others are arriving, exhausted, on the shores of Colombia. They don't know it, but in their feathers they carry crucial information about where they've been and what's happening in the habitats on which they depend, information that can help inform much-needed conservation efforts—if scientists like Keith Hobson and Camila Gómez can decode it.

Eight
The Feather Library

Ornithologists and wildlife managers increasingly recognize the importance of not only conserving diversity *across* species, making sure as many distinct species survive into the future as possible, but also conserving diversity *within* species—ensuring that each of the unique populations that make up a single species of warbler or flycatcher or hawk is preserved as well. And that means knowing where each of those genetically distinct populations is going when the birds leave their breeding grounds and head south.

You could say that the Bird Genoscape Project, the leading effort to map this variation in bird DNA, began with a conversation over coffee in 2009. Or you could say it got its start in Africa in the 1980s with some feathers taped into a notebook. Or you could argue that it really goes back twenty thousand years, to when massive ice sheets began to recede from North America for the last time.

One of the birds affected by that massive climatic shift was the Wilson's warbler, yet another member of the small, colorful, and much-loved wood warbler family that light-level geolocators have revealed so much about. Today, Wilson's warblers, small yellow birds with distinctive black caps, breed across Canada and parts of the

United States. But twenty thousand years ago, much of the habitat they now rely on would have been covered by a mile-thick sheet of ice.

At the height of the Last Glacial Maximum, ice covered a quarter of Earth's land surface area (compared with about 10 percent today), stretching as far south as the American Midwest. This was only the most recent of a series of repeating cycles of glacial expansion and retreat; Earth has experienced at least five ice ages. As the extent of the ice waxed and waned, the ranges of migratory bird species in the Americas likely contracted south and then expanded again, over and over. It's the signature of that final expansion beginning twenty thousand years ago that we can still detect in the DNA of birds like Wilson's warblers today.

During that final ice age, Wilson's warblers were forced to hug the eastern and western margins of North America; the central part of the continent, even south of the ice, was an inhospitable desert. The birds that nested to the east were separated from those that went west by enough distance, for enough time, that they developed genetic differences that can still be detected. And as Wilson's warblers followed the retreating ice into new regions, the new populations they founded were also genetically distinct from their forebearers. With each new generation, only a few trailblazing individuals would have moved into newly opened swaths of habitat, eventually filling them with descendants that shared their unique genes. Today, although Wilson's warblers that breed in the Pacific Northwest may *look* almost identical to, say, their cousins that nest in eastern Canada, each population carries genetic markers that tell the story of expansion after the end of the last ice age.

Hydrogen isotopes, after all, aren't the only source of information in bird feathers. There is also DNA, the molecules in our cells (and those of all living things) that carry the genetic programming passed down from one generation to the next. Like isotope studies, DNA analysis offers the promise of being able to pluck a single feather from a migrating bird and determine where in North America it started out. But while the ratios of isotopes in birds' feathers are a sort of useful accident, a by-product of the hydrology and geology of the places a bird happens to have been, DNA is the stuff of life itself.

All this means is that, in theory, you ought to be able to tell roughly where in North America an individual Wilson's warbler hails from by analyzing the contents of its DNA—sort of like how you can get an idea of where your own ancestors came from by swabbing the inside of your cheek and sending the sample off to a company like 23andMe. The evolutionary biologist Thomas Smith had an inkling more than three decades ago that something like this must be the case. But only recently has it really become possible.

In the late 1980s, Smith was a young PhD student gathering data on seedcrackers, small African finches with crimson faces and chunky gray bills. Working in the floodplains of Cameroon, he captured birds and measured their bills, tracked who was mating with whom, and even invented a tool for measuring the hardness of the seeds they ate. His work showed that black-bellied seedcrackers came in two distinct varieties, big-beaked birds that ate big, hard seeds and small-beaked birds that ate small, soft seeds, and he speculated that they could be on the path toward diverging into two different species.

Smith would later cofound the Bird Genoscape Project, and he told me that the beginnings of the idea go all the way back to that work in the 1980s. The genetic techniques available at the time were crude by today's standards, but while doing his field work, Smith started saving the occasional feather from the birds he was studying and taping them into the pages of his notebook. "I was taking blood samples, but I knew there was DNA in the feathers [as well]," Smith said in our interview. "I knew it was going to be difficult to get *out* of the feathers, but I thought, well, maybe it's like what Aldo Leopold said." Leopold, a renowned American conservationist and philosopher, wrote in 1949 that "to keep every cog and wheel is the first precaution of intelligent tinkering." Feathers were much, much easier to preserve and store than blood samples, so into the notebook they went.

After finishing his PhD, Smith accepted a faculty position at San

Francisco State University, where he started taking his ornithology students to the university's "field station" in the Sierra Nevada east of the city. There among the pines of Yuba Pass, Smith's students got to spend time helping with a bird banding operation, one of a network of bird banding sites across the country overseen by the nonprofit Institute for Bird Populations (IBP). Dubbed MAPS stations, for Monitoring Avian Productivity and Survivorship, these sites operate during the breeding season to collect data on birds' survival and breeding productivity. As his students helped extract black-headed grosbeaks and western tanagers from the nets, Smith was thinking about where those birds had been and where they were going. "[The banding station staff] were interested in using molecular techniques to see if they could connect bird populations in different ways," said Smith. "They thought, well, we'll collect some feathers and see what happens. And that's when I said, you know, we need to formalize this and start putting these in envelopes and archiving them and curating them. And so that was the start of the genoscape."

A "genoscape," the term Smith coined, is the genetics equivalent to Keith Hobson's "isoscape": a map of how genetically distinct bird populations are distributed across the landscape. Smith envisioned someday creating genoscape maps for migratory songbirds that would make the connections between an individual population's breeding and wintering grounds clearer than ever before. The research Smith was publishing in the 1990s was still mostly about birds' bill sizes and how they varied from one individual to the next, but on the side he began to amass an ever-increasing feather collection. It was Smith who, in the mid-1990s, provided many of the feathers Keith Hobson needed for his initial study proving that birds' movements could be traced back using deuterium.

In 2002, Smith switched jobs and moved to UCLA, where he founded the university's Center for Tropical Research and is now a distinguished professor in the Department of Ecology and Evolutionary Biology. Freezers full of feathers and tissue samples went with him in a rented semitrailer, speeding south along I-5. Their next stop was an infusion of new tech from one of the most massive international research collaborations of all time.

From Glaciers to Genomes

Today Kristen Ruegg is the co-director of the Bird Genoscape Project alongside Smith. When Ruegg joined Smith's lab at San Francisco State University in 2000 as a master's degree student, however, she didn't necessarily plan to build her whole career around birds, much less bird genetics. "I wanted to do research in the field of conservation biology," she said when I reached her via video call. "And so I actually came at it from a really practical [angle]. I knew that a lot of people really cared a lot about birds, and that if I worked on avian conservation, then it could have a trickle-down effect to conservation in general." But her path would soon come to intersect with the most famous genetics research initiative of the late twentieth century: the Human Genome Project.

For her master's thesis, Ruegg studied the Swainson's thrush—the same bird that Bill Cochran and Ana González followed on migration with two different generations of radio telemetry technology. Swainson's thrushes are split into two populations, one that breeds along the West Coast and one that breeds in central and eastern Canada, each with its own migration pattern. Ruegg's analysis showed that both groups quickly expanded their breeding ranges as the glaciers retreated, and that the odd migratory routes of some of the birds that nest in the center of the continent—which head thousands of miles east before turning south—are probably the result. These birds are flying back to the eastern breeding grounds of the ancestors, where their ancient programming kicks in and tells them to set out for South America.

Back in the 1980s, the primitive genetic techniques Thomas Smith used for his PhD let him look for differences in proteins between individuals but not for differences in DNA itself. Ruegg's work in the early years of the twenty-first century, however, focused on differences in the thrushes' mitochondrial DNA. Most of a cell's DNA is in its central nucleus, but a tiny bit is also housed in the mitochondria, the tiny organelles that generate the energy cells need to function. Mitochondria probably originated as independent single-celled organisms that, sometime in the distance past, were engulfed by and incorporated into bigger, more complex cells, and they still carry their own set of DNA, the code for proteins that are specific to the role mitochondria play in the body. But while a bird's nuclear DNA typically contains a billion or more base pairs, the individual bits of code that make up our genetic programming, its mitochondrial DNA might have only around twenty thousand. "Basically, it's simpler," said Ruegg. "It's just, like, super easy to sequence."

And while Ruegg was examining the mitochondrial genes of Swainson's warblers, another of Smith's master's degree students, Mari Kimura, was applying the technique to Wilson's warblers, digging deep into the feather library to extract the DNA of more than three hundred individuals captured across the species' breeding and winter ranges. It was the first real test of Smith's genoscape idea: Did the genes of Wilson's warblers vary enough across their breeding range that you could tell where in North America a bird had started out by plucking a feather from a bird in Mexico or Central America and sequencing its DNA?

Sort of. The birds' mitochondrial genes clearly sorted them into an eastern population and a western one, with the western population the more genetically diverse of the two. Western birds were more common than eastern birds at almost every winter site in the study. This wasn't the level of detail that would be really useful for figuring out whether, say, Wilson's warblers breeding in Alaska were more at risk from logging in Nicaragua than birds breeding in California. But it was a start.

In the following years, other collaborators of Smith's would begin to look for variations in small sections of Wilson's warblers' nuclear

DNA, even experimenting with combining genetic analysis with the stable isotope techniques developed by Keith Hobson (the idea being roughly that since the DNA varied between east and west, and deuterium varied along a gradient from north to south, together they might be able to better pinpoint birds' breeding locations). A 2006 grant from the NIH to study how migratory connectivity was linked to the spread of avian influenza among wild birds gave Smith's feather collecting a boost, cementing his partnership with the Institute for Bird Populations. The IBP trained its bird banders across the United States and Mexico in how to safely collect feather and blood samples from the birds they captured. But none of it was enough to really bring the genoscape concept to fruition.

Meanwhile, Ruegg moved on to a PhD program at UC Berkeley and then in 2010 to a postdoc at Stanford. It was while she was at Stanford, studying genetic markers in populations of whales rather than birds, that she took the class that provided the next break in the case.

"I was taking the courses for people in the human genetics world," she said, there being few courses available specifically on whale DNA. In one class, she learned that "there were all these competitions that had been created to incentivize companies to develop technology that would sequence things really quickly, to create the thousand-dollar genome." The advances covered in Ruegg's class largely stemmed from the massive international initiative to unlock the secrets of human DNA: the Human Genome Project.

Formally launched in 1990 and completed in 2003, the Human Genome Project was a government-funded initiative to sequence an entire human genome—that is, to figure out the exact order of all of the base pairs that make up the entirety of a single person's nuclear DNA. A human genome contains about three billion base pairs (three times as many as a typical bird's), so this was a massive undertaking. The entire effort, from start to finish, cost around three billion dollars, and the results led to huge leaps in our understanding of the genetic basis of many diseases, the evolution of humans from earlier hominids, and many other areas.

As part of the race to decode the human genome, the scientists involved developed new techniques that have revolutionized their field. Crucially, they invented what have become known as "high throughput" sequencing techniques. Decoding DNA requires a tedious process of copying and chopping up extracted DNA, drawing the resulting bits through a slab of gel using electrical current, and identifying the individual molecules that make up DNA sequences by how far and how fast they travel. Although the basic principles haven't changed since Thomas Smith was a PhD student, Human Genome Project scientists, spurred on by government grants and incentives, found new ways to automate this process.

So while Smith and his collaborators were painstakingly examining a handful of locations in the nuclear DNA of Wilson's warblers to see how they varied from one individual to the next, the researchers at the Human Genome Project had figured out how to sequence the large volumes of DNA they were dealing with quickly and cheaply. Essentially, they had made the field of *genomics*—studying the genome as a whole instead of focusing on individual chunks—more accessible than ever before. It took time for these breakthroughs to make their way to the world of zoology. But sitting in that Stanford classroom, Ruegg remembered her old mentor's dreams.

After finishing the class, Ruegg traveled to Los Angeles to visit some relatives. While she was there, she met up with Smith for coffee. Musing on the next step for her career, "I was telling him that I was interested in maybe trying soft money," funding her own research from sources outside the university system, according to Ruegg. "He said, okay, great, we should write grants together. And I was like, hey, you know that idea you had with the bird genoscapes, tracking migratory movements using genetics? I think we could really do that now."

Smith persuaded a wealthy former board member of the National Audubon Society to chip in enough money to cover Ruegg's salary for the first year, and the Bird Genoscape Project was officially born.

Managing the Freezer Farm

Behind a locked door in the basement of a building on the Colorado State University campus in Fort Collins, where Kristen Ruegg is now an assistant professor, is the freezer farm. Row after row of freezers stores precious samples for the scientific work being carried out on the floors above. In four of those freezers—ordinary appliances originally purchased at Lowe's—are the Colorado component of the Bird Genoscape Project's feather collection.

Christine Rayne, Kristen Ruegg's lab manager, pulls open a freezer door. "We've just bought two new ones, because we're getting full," she says.

Rayne was giving me a virtual tour of the Bird Genoscape Project's facilities at Colorado State. While Ruegg and Smith dream up new projects, write grants, and edit papers, Rayne is the one who keeps things running day to day, and I wanted to learn exactly what that entails. Wearing a black COVID face mask and carrying an open laptop with our Zoom call running in one hand, she walked from floor to floor of the building, showing me the lab space and the equipment they use for extracting and analyzing DNA and, finally, the samples themselves.

In 2019, Rayne was wrapping up her bachelor's degree at Colorado State and trying to figure out what to do next. She was interested in genetics, but unsure whether she wanted to go to med school or vet school or apply for a PhD program or something else. "One of my professors told me, oh, you're interested in genetics and conservation? There's a new professor in the biology department, she's just starting, you should talk to her," said Rayne. Rayne set up a meeting with Ruegg and ended up volunteering in her lab for the rest of the year while finishing her degree and working part-time at a veterinarian's office.

"After that I kind of did the ski bum thing for a while, which was great, but then Kristen reached out and said, hey, our lab manager's leaving, are you interested in applying?"

Rayne is the person who does much of the hands-on work involved in keeping the Bird Genoscape Project running. She manages the Colorado portion of the vast sample collection—192,144 samples in the database the day we talked, split across Smith's lab in California and Ruegg's in Colorado, including blood and tissue as well as feathers. She supervises undergraduates, the role she herself was in just a few years ago, who assist with properly cataloging and preparing new samples. She extracts DNA, doing some types of analyses herself in-house but often preparing the extracted and concentrated DNA to be shipped off to private companies that do whole-genome sequencing. And she also manages budgets, orders supplies, applies for permits, and takes care of all the other little tasks necessary to keep a research lab running.

"We're constantly receiving samples, sometimes huge collections from people," said Rayne. "Some people send us a really, really nice Excel spreadsheet of everything that they sent, and it's nicely cataloged, and some people just send us a bucketful of feathers." Project staff fell behind on entering new samples into the system during COVID, but Rayne, who started her job in the spring of 2021, was working to catch up.

After showing me where the undergrads sit as they label feather envelopes and type the information into the database, she carried her laptop into the "clean room" where the actual DNA extraction happens, a space kept as isolated from the outside world as possible to minimize the chances of contaminating the genetic material. Extracting DNA from blood is simpler than getting it out of feathers, just because there's more genetic material to work with, so many studies begin with blood and then move on to feathers, which are easier to collect in the field. Getting DNA out of a feather is a two-day process that begins with carefully cutting off a feather's calamus, the bare, hollow bit at the base of the feather shaft, where a bit of DNA-containing tissue is lodged at the point where the feather attached to the skin. (Feather cutting "doesn't really require skill, and it's time consuming," making it another perfect job for the undergrads, she says.) The calamus goes into a test tube where a series of chemicals are used to completely dissolve it, lysing, or breaking down, the individual cells. Finally, the resulting liquid is spun in a centrifuge and filtered to isolate and concentrate the bits of DNA it contains.

Leaving the extraction room, Rayne carried her laptop down several flights of stairs to our last stop: the basement freezer farm. The freezers for all the various research labs housed in the building were all in the basement, she explained, because that was the only part of the building with backup power in the event of an outage. The blood and tissue samples were in one basement room, and the four feather freezers in another. The shelves of each feather freezer are lined with neat rows of plastic tubs, each tub stuffed with the small envelopes that store the individual feathers. Rayne pulled out one labeled "American Robin" to show me. "Each feather gets assigned a unique ID number when it gets entered into the database," she said, "so if I was looking for AMRO 05N 2464, I could look for the AMRO 05N box, and then all of the feathers here are just in numerical order."

There in the basement, box after box after box of feathers waits for its chance to contribute to science. But these feathers have often passed through many hands before they get to Rayne. Rafael Rueda Hernández, the Mexico coordinator for the Bird Genoscape Project, sources feathers from bird banders in Mexico and gets them safely and legally to the lab in Colorado. "Most of the [migratory] bird breeding populations in North America, they at least have to go through Mexico at some point of their life cycles," he said when I called him to learn more about this end of things. "It's a very important area of interest for full-cycle ecology and for migratory connectivity studies." He works with Institute for Bird Populations banding stations as well as independent banders and researchers. "My job is to keep them interested, you know, keep them sending feathers and contributing to the project."

Like Rayne, he also does a lot of paperwork. First he needs a permit to collect feathers and other samples from wild birds, then a permit to assure that the samples aren't biologically hazardous material, and then a third permit to be allowed to export them—and that's just on the Mexican side. The Mexican permits then enable him to get a U.S. Fish and Wildlife Service import permit and a USDA permit. Governments have put these steps in place to try to prevent the illicit trade of endangered species and the spread of wildlife diseases, but when you list it all out, it sounds almost comical.

Hernández sometimes ships feathers via FedEx or another delivery service, but according to him, shipping companies' Mexican staff are often not fluent in English and are thrown off by pages and pages of English-language permits peppered with scientific names. "It has happened that the samples get lost," he said. So when he can, he transports samples himself, bringing them with him through customs when he flies into the United States. "We have to send an email to the U.S. Fish and Wildlife Service and let them know that we are entering the country with biological material," he explained, and someone from the U.S. Fish and Wildlife Service will meet him at customs to inspect the feathers or blood samples he's carrying. "The first time, I was anxious," he said. "As a Mexican [entering customs], you're on the spot, and now you're a Mexican with biological material. So that's hard."

Hernández's favorite part of his job is doing outreach: he never turns down an invitation to give a talk about the Bird Genoscape Project in Spanish, traveling all over Mexico to give talks to bird banders and other potential collaborators. "All the papers, all the main results are in English," he said. "And those results are what keep people interested. So I try to close that gap of knowledge about the project."

Rayne and Hernández are just two of the dozens of students, staff, and postdocs who have contributed to the Bird Genoscape Project over the years, contributing innumerable hours of labor collecting, transporting, and analyzing the tens of thousands of samples that make up the feather library. Today the collection includes samples from hundreds of species. But the first official genoscape focused on the bird that Thomas Smith and his colleagues had been eyeing since that first mitochondrial DNA study: the Wilson's warbler.

"Completely Awesome . . . So Psyched"

It took five years to get from Smith and Ruegg's fateful coffee meeting in Los Angeles to the debut of the first full genoscape. During those five years, Ruegg came perilously close to abandoning her research career entirely. But when the result—which focused on the Wilson's warbler, the same broadly distributed warbler species that had previously been studied using mitochondrial DNA and earlier nuclear DNA techniques—was finally published, it was a blockbuster.

The method they used to tease out the fine-scale variations in Wilson's warblers' nuclear DNA required a lot of work. Beginning with a pool of about 450,000 individual sites in the warblers' genomes, they compared pair after pair of birds until they winnowed this down to just 96 key spots that were most likely to differ between warblers from different parts of the continent. Then they analyzed the variation at those key spots in DNA samples from more than 1,600 birds captured at 68 different locations.

"I'd say it was three years of gathering and analyzing the data," said Ruegg. Once that was done, they still had to write the manuscript, and then there was the tedious process of formatting it, submitting it to a scientific journal, and waiting for it to go through the formal review process. In our conversation, Ruegg was clear that she always loved the work itself, but the pressure of balancing her own career in science with the needs of her family nearly derailed her.

She was living in Santa Cruz, California, where her husband had a job with the federal government, and caring for her young daughter while trying to string together grants to keep working on the project. "I was barely making enough to cover day care," she said. "It was ridiculous. Part of the reason it took so long is that I was pretty much a full-time stay-at-home mom while also doing the bird genoscape work on very minimal funding." She found herself seriously considering returning to graduate school for a degree in science communication and attempting to pivot into a new career.

But when the results finally came together, it was clear that her efforts had paid off. The blurry picture of Wilson's warbler diversity that had been drawn from mitochondrial DNA and limited nuclear DNA studies finally snapped into focus. "It was completely awesome," said Ruegg during our conversation, grinning as she recalled the heady feeling of seeing the full genoscape laid out in front of her for the first time. "We were making blowup maps and putting them up on the wall. We were so psyched. It was crazy."

With this level of genetic detail, Ruegg, Smith, and their colleagues could distinguish not only eastern and western Wilson's warblers but *six* genetically distinct groups of birds with different migratory routes and different wintering grounds. And once the procedures had been worked out, telling which of these populations an individual warbler belonged to was cheap and easy: a well-equipped lab could get through three hundred feather samples in a day, for just ten dollars each.

"One of the central challenges in migrant conservation," the researchers wrote, "is that population declines and conservation planning often occur at regional spatial scales, but our knowledge of population structure and migratory connectivity is often limited to species-wide range maps." The Wilson's warbler genoscape showed that the Sierra Nevada, coastal California, and Pacific Northwest breeding populations all overlapped during the winter on the Baja peninsula. However, regional surveys showed the Sierra Nevada Wilson's warbler population was declining, while the other two were holding steady. Whatever was causing problems for the Sierra Nevada warblers, it appeared, must be tied to their breeding range, rather than where they spent the winter. This is the sort of detail that can help wildlife managers pinpoint conservation efforts and prevent individual populations from blinking out.

The researchers could even look at the timing with which birds from different populations passed through stopover sites. A banding station in southwestern Arizona collected feathers from Wilson's warblers passing through on their way from the beginning of March to the end of April. Extracting DNA from the feathers and identifying the population to which each bird belonged revealed distinct waves of migrants: birds from the coast of California passed through first, followed by birds from the Pacific Northwest and Sierra Nevada, with warblers that had bred farther north in Canada and Alaska finally straggling through in the last weeks of April.

Ruegg presented the results at an ornithology conference in Estes Park, Colorado, in September 2014. On a crisp Thursday morning, in a YMCA auditorium at the foot of the Rocky Mountains, she gave a fifteen-minute talk titled "Mapping Migration in a Songbird Using High-Resolution Genetic Tags." (Keith Hobson, coincidentally, had the slot in the program immediately after her, talking about the relationship between stable isotopes and geolocators.) It was a hit.

"Everybody saw it. Everybody noticed it," she said. Over the remaining two days of presentations, other researchers were mentioning it in their *own* talks. By the staid standards of ornithology conventions, it was a sensation. "It was definitely the biggest breakthrough that I've had in my career to date for something where I personally did the vast majority of the data collection and analysis. Now I get to watch students make these awesome discoveries, and that's really fun, too . . . but I haven't had one like that for a while." The momentum from the publication of the Wilson's warbler genoscape led directly to her current faculty position at Colorado State University.

The Bird Genoscape Project's goal is to complete genoscape maps for one hundred migratory species. By 2021, they had completed about twenty, with funding for thirty more. The most recent one published, for the American kestrel (the same small falcon whose claws Sadie Ranck was analyzing for deuterium in Boise), identified five unique kestrel populations spread across the United States and Canada, previously unrecognized divisions within the species that can serve as the basis for future conservation planning. For that paper, Ruegg, Smith, and their colleagues skipped over the process of identifying the most important individual sites of variation in the kestrels' DNA and jumped straight to sequencing their entire genomes, because at this point the technology has advanced so much that that's actually the cheaper, easier option.

After we spoke, I emailed Ruegg to make sure she was comfortable with me including details about her early struggles in this book. Her answer was an emphatic yes. "I think it is so important for people, especially moms, to know how hard science can be to juggle with all the other demands they may have on their time at these critical points," she wrote. "I have always found my most creative moments strike when I am one step away from giving up altogether. It's almost as if I say to myself, well, what do I have to lose? Let's go for it!"

A Climate Crystal Ball

Genoscapes can tell us in incredible detail what migratory bird populations look like now. But in some cases, they can also give us a hint about the future.

The main focus of the Bird Genoscape Project is migratory connectivity: which individual populations within migratory bird species go where, and at what points in their annual journeys they face the greatest threats. But one former postdoc on the project, Rachel Bay, came from the world of coral reefs, where the conservation needs are different. Coral reef scientists are experimenting with using genetics to predict which individual corals will be most susceptible to large-scale threats such as bleaching events and climate change. What if those same techniques could be applied to birds?

The species Bay focused on for her postdoctoral research was the yellow warbler, a sunny yellow bird whose cheerful song is often written as "sweet, sweet, a little more sweet!" They're the most common warbler species around my home in eastern Washington in spring and summer, but I share my enjoyment of them with birders across the continent: they have a vast breeding range that extends from coast to coast across North America and from Alaska down into Mexico. These birds experience drastically different habitats during their breeding seasons, from the cool, rainy coasts of the Pacific Northwest to the dry heat of California's Central Valley to the thunderstorms and mugginess of summer in the Midwest. It seemed reasonable to imagine that these different yellow warbler populations might have genetic adaptations to the specific climates they called home—which might leave some more vulnerable to climate change than others.

Thomas Smith's feather library, boosted by additional contributions of yellow warbler feathers from the Institute for Bird Populations and other sources, gave Bay and her Bird Genoscape Project colleagues the raw material they needed to find out if this was true. They looked at the genomes of 229 individual yellow warblers from twenty-one different sites across their breeding range. "DNA is a string of As, Ts, Cs, and Gs," Bay told me. "Let's say we have a bunch of birds in the Central Valley, here where it's hot and dry during the summer, and at a particular location in the genome, they all have As. And then let's say we compare that to a population of birds that are somewhere cold and wet, and at that same place in the genome, they all have Ts. Our first step in this process would be to say, okay, at this location in the genome, we have a statistical association between climate and the base pair that we have here."

Once Bay had untangled the links between the birds' genes and the climates they live in today, she could compare that with the ways those local environments are going to change under likely climate change scenarios. Let's say a cool, wet place becomes a little hotter and drier; to thrive there in the future, more of the yellow warblers in the local population would need to have As at the relevant spot in their genome. "Genomic vulnerability," as defined by Bay, is that difference between what genes are present in a bird population now and what we predict they'll need to have in the future. The bigger the gap, the more a population is at risk as the climate changes. And this isn't just theoretical: Bay's work showed that yellow warbler populations with the greatest mismatch are already declining faster than populations that are a better genetic "fit" for the changes predicted for their region.

Of course, migratory songbirds like yellow warblers spend less than half the year in their breeding range. In a follow-up study, Bay showed that, fascinatingly, individual yellow warblers seek out similar climates year-round: birds that spend the summer in drier parts of North America also spend the winter in drier parts of Central and South America. "I was super excited about that result, because we've been thinking about local adaptation only in the breeding range," said Bay. "There are many different ways that analysis could go; the climate in the breeding range could be completely random with respect to the wintering range. And that would make natural selection total chaos, right? So it's interesting that there's this correlation, because I think it provides a way for natural selection to reinforce itself across the annual life cycle, and for birds to be more likely to be actually adapted to their specific local environments."

According to Bay, we could help the most at-risk populations by removing as many other stressors as possible, minimizing threats like habitat loss and predation so that birds at least don't have to contend with those on top of a changing climate. But translating work like this into on-the-ground conservation actions will take time. Wildlife managers and policy makers are unlikely to make a decision based on the vulnerability of a particular yellow warbler population, so for data like this to have an impact, we'll need similar studies from multiple species.

In 2020, a team of Bird Genoscape Project scientists led by Kristen Ruegg published a study looking at how to use genoscape data to prioritize limited conservation funding to best benefit their old friend the Wilson's warbler. Ruegg works closely with conservation nonprofits like the National Audubon Society and federal agencies like the U.S. Fish and Wildlife Service to try to make sure her data is put to good use, but hard choices will have to be made; her work showed that the best way to spend conservation dollars differs depending on whether you want to preserve the greatest number of individuals or the greatest amount of genetic diversity within a species.

"I'm biased, but in my opinion, you know, everything comes back to genetics," said Ruegg. To some extent, some bird populations can respond to a changing climate by moving into new areas, but that won't be enough to sustain many species in the long term. "They're either going to have to adapt or not, and so understanding the adaptive capacity of a population and the traits that are important to local adaptation is really critical not only for being able to forecast how populations will respond to climate change but also for developing management strategies that might include assisted migration or captive breeding."

Insights about bird migration gleaned from genomics—and from stable isotopes, and weather radar, and the many other techniques covered in this book—are invaluable for understanding the species we love and ensuring they'll still be here for us to enjoy in the future. But they're most useful when combined with information from the people who are out in the forests and fields with their binoculars, observing where birds are and what they're doing on any given day. Remember Bay's finding that yellow warbler populations with the greatest genomic vulnerability to climate change are the ones already declining fastest? That data on where yellow warblers are becoming more scarce came from the North American Breeding Bird Survey, an annual volunteer effort in which bird-watchers comb established roadside routes in spring and report what they find.

The term for this type of volunteer data collection is "citizen science"—or, more recently, "community science." And although tiny transmitters, mass spectrometers, and Doppler radar stations are great, if you have a smartphone in your pocket, you already have all the equipment you need to join the cause of studying and conserving migratory birds.

Nine
Vox Populi

It's a cold Friday afternoon in February when I park outside a locked gate at the edge of Walla Walla, Washington. As I get out of my car, making sure the strap of my binoculars isn't twisted, the wind tugs at my hair. I should have brought a hat.

On the other side of the dirt driveway, Bruce Toews is also getting out of his car. We've never met, but I've persuaded him via email to let me join him for his end-of-the-week bird walk around Bennington Lake. Since 2020, Toews has completed more than seven hundred outings like this one, meticulously recording every bird he sees and logging them in a database used by ornithologists and wildlife managers. Toews himself, however, is not an ornithologist. He's a professor of accounting and finance at the Seventh-Day Adventist university in the next town over.

The reasons that drive Toews and thousands of others like him (including, on occasion, myself) to spend their spare time collecting data on birds are probably as varied as the people who participate. There is, however, a term for what they (we?) are doing: community science.

There are several definitions out there for the phrase "community science," but essentially it refers to projects where people who aren't professional scientists contribute to scientific research, usually by collecting data to pass on to researchers. (This is also widely known as "citizen science," a term independently coined by multiple sources in the mid-1990s. However, in recent years many organizations have been moving away from this term due to the political connotations related to the concept of who is and is not a "citizen" of a particular place. I will be using the modern phrase "community science" throughout this chapter.)

This type of data collection has a long history in ornithology, dating back to before either of the terms in use today was coined. In the early days of North American ornithology, the line between professional scientist and amateur bird enthusiast was a blurry one. It was common for people with "day jobs" as doctors or lawyers to also record their observations of birds' characteristics and behavior and publish their findings in ornithological journals. "Until comparatively recently," according to one history of the subject, "virtually all of American ornithology was the product of amateurs."

Over the course of the nineteenth century, ornithology gradually became recognized as a profession in its own right, but it remained common for professionals and amateurs to work together to collect data. George Lowery's amazing continent-wide moon-watching project in 1952, for which he and Bob Newman recruited thousands of volunteer observers, would today be considered an example of community science; so would the contributions of the amateur nocturnal flight call enthusiasts who record and share evidence of the birds that pass over their homes at night.

The longest-running community science project in the world, however, is the venerable Christmas Bird Count (or CBC), which has been happening every year since 1900. Its history is almost a legend among birders: In the nineteenth century, the story goes, Christmas "side hunts," where hunters would head out on Christmas Day to see who could shoot the most birds, were a popular tradition. A conservation-minded ornithologist named Frank Chapman, however, proposed that it might be more humane and contribute more to our understanding of bird populations if people merely *counted* the birds they saw instead. In 1900, he and his colleagues at the fledgling Audubon Society organized the first Christmas Bird Count, recruiting twenty-seven participants in 25 locations in Canada and the United States and tallying a total of 89 species. In December 2019, the 120th year of the count, more than eighty-one thousand bird-watchers participated in 2,646 count events throughout North America, Latin America, the Caribbean, and the Pacific Islands, collecting observations of forty-two million birds representing 2,566 species.

I've joined a few counts myself over the years, spending the day cruising back roads in chilly rural Ohio or walking along a saguaro-lined trail in southern Arizona with a small group of other bird enthusiasts, our eyes and ears peeled, tallying every bird we can identify. Modern Christmas Bird Counts are organized around "count circles" fifteen miles in diameter. Participants surveying a specific circle split off into small groups to cover as much ground as possible, then traditionally reconvene in the evening for potluck "compilation dinners" where they add up totals of each species seen in the circle that day.

Data collected on Christmas Bird Counts is useful for estimating how populations of birds increase, decrease, and shift their ranges over time. I'd long assumed, though, that because of when the counts take place—during the winter, when birds that migrate between North America and the tropics are largely absent from the United States, where most counts still take place—Christmas Bird Count data probably didn't have much to tell us about migration. When I said as much in a tweet, however, the ornithologist Emily Williams posted a reply to set me straight.

Williams, a PhD student at Georgetown University, is studying the migration of American robins, whose seasonal movements back and forth within North America vary widely depending on where they live and other factors; robins that breed in Alaska may fly thousands of miles south to escape the north's frigid winters, while some of their cousins at more southern latitudes stay put year-round. Williams has been putting GPS tags and other tracking devices on robins around the United States to tease out the details of these patterns, but she also plans to dig into Christmas Bird Count data to see if their migration habits have changed over time. "I've been talking with people to try to get a sense of where large concentrations of robins are hanging out, to try to select study sites," she told me when I reached her for a video chat, "and I'm hearing from a lot of people, especially in more northern locations, who'll say, oh yeah, I've been doing the CBC in this area for twenty years, and twenty years ago we never saw robins, and now we see thousands."

Williams acknowledges that CBC data is less useful for studying the migration of species that leave North America for part of their annual cycle. But since the inception of the Christmas Bird Count, a host of other community science programs focused on birds have been launched. You can help track the northward migration of hummingbirds in the spring for Journey North, monitor avian visitors to your own backyard for Project FeederWatch or the Great Backyard Bird Count, listen for the calls of whip-poor-wills for the Nightjar Survey Network, or comb beaches for shorebirds for the International Shorebird Survey, to name only a few. One spring, when I was eight months pregnant, my husband and I volunteered for a program to monitor short-eared owl populations, spending several evenings driving a route east of town at dusk and pausing at regular intervals to watch for the big pale birds coursing back and forth over the wheat fields. (We never spotted any.) And community science today goes far beyond birds as well. A smartphone app called iNaturalist will let you document your observations of any living species, from anywhere in the world. If the outdoors isn't your thing, you can even play a computer game to help crack new ways of folding proteins or donate your PC's spare processing power to analyze radio signals from outer space for signs of intelligent life.

There are two programs, however, that arguably have the most to offer when it comes to improving our understanding of how migratory bird populations are connected. From the birth of the modern environmental movement in the 1960s, to the age of smartphones and machine learning, community scientists have often been the ones who've made it possible to know how many migratory birds there are and where they go and when.

The Dow Jones of the Bird World

"Birds both inspire and permit citizen science; we can find them, we can identify them, and we can count them. And many people will happily get up early to do so."

These words appeared in the introduction to a special section in the ornithological journal *The Condor*, published in 2017 to commemorate fifty years of the North American Breeding Bird Survey, or BBS. The Christmas Bird Count had been chugging along for six decades by the time the BBS, an annual breeding-season survey that's provided crucial information on how migratory bird populations are faring over time, was initiated in 1966. But it was a visionary colleague of Rachel Carson's who fully realized the potential of volunteer bird-watchers to fuel large-scale insights about what was going on with North America's birds.

Chandler "Chan" Robbins was born in Massachusetts in 1918. His parents and grandparents were all interested in the natural world, and one of his earliest memories was passing by a display of mounted birds at the local library while being pushed in a stroller as a toddler. By the time he was sixteen, he was already organizing and compiling his own local Christmas Bird Count. (According to an obituary of Robbins, his first-ever CBC was marked by controversy when an experienced ornithologist questioned his report of Barrow's goldeneyes, a duck species not typically found in the area, at a local pond. Robbins, however, was ultimately vindicated.)

In 1940, he graduated with a bachelor's degree in physics from Harvard University, where his mentors warned him about the lack of career opportunities in his original field of choice, ornithology. He briefly taught high school science in Vermont, but when the United States entered World War II, his career took a different turn.

Influenced by an uncle who'd served in World War I, concluded that war was morally wrong, and become a minister, Robbins declared himself a conscientious objector. As an alternative to military service, he joined a program called the Civilian Public Service and eventually obtained an assignment at the U.S. Fish and Wildlife Service's Patuxent Research Refuge in Maryland. When the war ended in 1945, he was hired on permanently to do ornithology work for the USFWS. He would work as a biologist for the federal government for the next sixty years.

Multiple threads from Robbins's first decades at the USFWS would eventually come together to spur the birth of the BBS. One of his many jobs in those early years was designing roadside surveys for game bird species such as woodcocks and doves that would make it possible to keep track of whether their populations were increasing or decreasing. He also expanded his involvement in the community science initiatives of the era (though it was still decades before the terms "citizen science" or "community science" would be coined), overseeing a handwritten database of volunteer-submitted records of bird phenology (the timing of annual events in their life cycles, including migration) stretching back to the nineteenth century as well as continuing his involvement in the Christmas Bird Count. Finally, and crucially, he studied the effects of the notorious pesticide DDT on wild birds, eventually working with Rachel Carson. Carson edited Robbins's reports on the subject before going on to write *Silent Spring*, the book that helped kick off the modern environmentalist movement. Robbins, in turn, used *Silent Spring* in his lobbying efforts to get USFWS leaders to take declining bird populations more seriously.

Around 1962, Robbins would later recall, he got a letter from someone in the Midwest who was concerned about large numbers of robins being killed by pesticide spraying, inquiring about whether this was affecting the robin population on a national level. "I had to write her

back and say, at the moment we don't have any way to measure continental populations of birds," said Robbins in a 2008 interview. "There
were waterfowl surveys . . . but we didn't have any civil programs to
measure what's happening with songbirds."

Robbins believed that the same roadside protocols he'd developed
to survey game bird species could be used to monitor songbird populations across the continent, if enough volunteers could be recruited to
carry out the surveys. In 1966, after a year of pilot projects in Maryland and Delaware, Robbins officially launched the national Breeding Bird Survey.

It wasn't easy. Many government biologists in the 1960s still saw
little point in collecting large amounts of data on nongame species,
and Robbins had to hustle to fit his work on the BBS in around what
his superiors saw as the more important parts of his job, keeping
the program running with the bare minimum of time, resources,
and money. Many were also skeptical of the methodology that the
BBS relied on, questioning whether surveys conducted along roads
were really representative of the broader landscape. His own wife
was apparently dubious about whether amateur bird-watchers could
be persuaded to follow the standardized protocol that would be required.

But Robbins persisted. In 1966, volunteers carried out six hundred surveys in the eastern United States and Canada. Just two years
later, in 1968, the BBS had already ballooned to include two thousand
routes across the full breadth of southern Canada and the contiguous
United States. (Today that number is more than four thousand.) BBS
data couldn't provide estimates of the total population sizes of North
America's bird species, but crucially, because it was taken at the same
places year after year, it could give an idea of *trends*, whether those
populations were increasing, decreasing, or holding steady over time.

To recruit volunteers, Robbins worked the network of professional
and amateur bird enthusiasts he'd cultivated over his years in the field.
Christmas Bird Count participants, local bird-watching clubs, and
bird banders all provided pools of potential surveyors. Today a little

more than half of the BBS's volunteer surveyors work in conservation-related jobs; the rest are enthusiastic, highly skilled amateurs. The only requirements are access to a reliable vehicle, good eyesight and hearing, and the ability to identify all breeding birds in a region by their songs and calls as well as by sight. The basic procedure that volunteers follow hasn't changed since Robbins developed it in the 1960s: Surveys are carried out in June, at the height of the breeding season, along set routes twenty-four and a half miles long. Beginning half an hour before sunrise, the surveyor pulls over their car every half mile and records every bird they can see or hear in three minutes. Today volunteers can submit their data online once it's collected, but even in the internet age they still use paper data sheets in the field, mailing them into the BBS's overseers at the U.S. Geological Survey (which eventually took over the program from the USFWS) to provide a permanent physical record.

One of these people who rises in the wee hours of the morning to spend five hours painstakingly recording every chirp and flash of feathers along their assigned route is Laura Steadman, a software engineer in Boulder, Colorado, whom I met after putting out a call on Twitter for BBS volunteers who might be willing to chat with me. Steadman originally planned to pursue a career in archaeology, and in 2010, while doing an internship with the National Park Service, she was unexpectedly offered a chance to travel to the Gulf Coast to be part of the federal government's response to the Deepwater Horizon oil spill. There, she helped ensure that sea turtle nests and dune vegetation weren't disturbed during the ongoing cleanup efforts.

"We were supposed to keep an eye out for migratory shorebirds," she told me during our video call, "but they all looked the same, and I struggled to care much." But one night, while she was driving down the beach in a golf cart, her lights caught a bird unlike anything she'd ever seen before. "It had this huge bill. It looked like a toucan. I was like, who do I have to call about this bird? Because this is weird, it can't be normal, it has to be some endangered thing." Flipping through the little guide she carried, however, she realized she was looking at a black skimmer, a common (though undeniably odd-looking) beach bird in that part of the world.

"That really floored me," said Steadman. "This is a crazy bird that I didn't know existed in this world, and it's totally normal for it to be here, it's just that I haven't been looking for it."

Steadman was hooked. She was officially a birder.

Eventually she moved to Colorado, where, unable to make any headway in the archaeology field, she attended a coding boot camp and eventually found work as a software developer. Birds, however, remained a constant in her life, and she was active on a listserv dedicated to Colorado bird-watching. "I'd seen posts through the years, occasional calls, like hey, we've got a BBS route that we need someone to run," she said. At first she felt as if she didn't have the skills necessary to do the Breeding Bird Survey justice. But over time, "I started to feel more confident in my skills, and eventually I replied to one of those emails and said, I'm interested."

Steadman's first survey was in 2019. Her route is located three hours away from her home in Boulder, so she dedicates an entire June weekend to it. She drives out the day before to briefly scout out her assigned route, making sure her stops from the previous year are all still accessible, then camps at a state park near the starting point.

"My official start time, according to my section paperwork, is 4:52 a.m.," said Steadman. She sets her alarm for 4:00 and drives the thirty minutes back to the start of her route in darkness. Once she arrives, she cracks open a bottled coffee and gets out her stopwatch, her binoculars, and her clipboard with her data sheets. "Depending on the year, the weather could be really great, but last year it was really cold and damp. You do a three-minute point count, hop back in the car, drive a half mile, do a three-minute point count, and you do that for fifty stops. It takes like four to five hours, so you're done around ten." Her route is in a sparsely populated area of the Great Plains where meadowlarks, horned larks, and lark buntings are among the most common songbirds.

I wanted to know why she remains committed to the BBS despite the significant time commitment. "I love it," she says. "It validates my skills as an amateur birder that I feel comfortable and confident to do it. It feels like an honor to be able to contribute to this data set as a citizen. It's so long running, and it's taught us a lot about bird populations and migration, so that feels really cool."

But really, it's just about getting out there with the birds. "I think one of my favorite reasons to keep doing the BBS, what keeps me going back out there at 4:00 a.m., is that predawn chorus that you get to hear," she said. That moment just before the sun crests the horizon when the noise of meadowlarks and other birds singing from every direction becomes almost overwhelming—for Steadman, there's nothing else like it.

To get a better idea of how the BBS has changed over the years and where it stands today, I contacted Dave Ziolkowski, the USGS's current program coordinator for the survey. Ziolkowski first encountered Chan Robbins when, as a teenage bird-watcher in Maryland, he was hired to help with bird surveys of one of the last remaining stands of old-growth forest on the Atlantic coast. He was to meet the biologist he'd be working with, a man in his mid-seventies, at the site early in the morning.

"It was a very hot day in June," Ziolkowski told me, "and I remember thinking to myself, I'll have to go easy on this fella." But when Ziolkowski arrived at the spot where they were to do their first bird count, "I saw him going through a thick, thick thicket of greenbrier, which has very thick thorns, and he just dropped through it like it was nothing. And I realized, holy smokes, this is going to be a long day." That aging ornithologist, of course, was Chan Robbins.

Robbins had continued to work hard throughout the intervening decades to oversee and expand the BBS and prove its value. In the late 1980s, he and his colleagues published a paper that sounded the first alarm about the ongoing population declines that have worried ornithologists ever since: BBS data clearly showed that after a period of stability migratory songbird populations had begun to drop around 1978. "That's what really brought this concept of conservation of migratory songbirds to the forefront for many people," according to Ziolkowski.

But in addition to showing what's happening to North American bird species at continent-wide scales, BBS data can be broken down by region—which is part of what makes it so useful for migratory connectivity research, looking at which populations within a species migrate to which locations in the Neotropics and whether some populations might be more vulnerable than others as a result of their specific journeys. It was BBS data that let Kristen Ruegg determine which of the six Wilson's warbler populations in her genoscape were increasing and which were decreasing. BBS data helped make sense of Rachel Bay's studies of yellow warblers and climate change, and of Gunnar Kramer's tracking of golden-winged and blue-winged warblers with geolocators. "The BBS essentially provides the Dow Jones of the bird world," Ziolkowski told me. "So we are not a total count of everything that's happening in the bird world, but we are a specific index that tracks the abundance of birds over time at very broad scales."

I asked Ziolkowski if he thought Chan Robbins could have envisioned the level of detail we'd someday have on these birds' migratory movements when he started the BBS in the 1960s. Ziolkowski said he'd asked Robbins that himself, and the answer was a frank no, never. But that's hardly surprising, since Robbins's career as a federal biologist spanned six adventure-packed decades. The more research I did about him, the more fascinated with him I became.

For example, if you're into birds and spend much time on the internet, you might have heard of a famous Laysan albatross named Wisdom. Originally banded on Midway Atoll in 1956, she has (as far as we know) continued to return to the island to lay an egg each December for at least seventy years, making her the oldest confirmed wild bird in the world. Care to guess who originally banded her? Yup— Chan Robbins, a fact that, when I read it, caused me to get up from my desk, walk into the next room where my husband was working, and exclaim, "Did you know that *Chandler Robbins is the person who banded Wisdom*!" Even more astonishingly, it was Robbins who rediscovered

Wisdom in 2001, thus documenting her longevity. Still working for the federal government, he returned to the atoll in his early eighties and happened to be the one to recapture her for the first time since 1956. USFWS scientists have eagerly anticipated her return every year since.

Even after Ziolkowski completed his graduate studies and came to work at the USGS, "I was always just sort of that little kid who was still learning from [Robbins]," he said. "And he had an amazing memory and could recount pretty much everything he had done in his career, even specific birds on BBS routes from decades ago."

In a 2008 interview conducted by the U.S. Fish and Wildlife Service as part of an oral history project, Robbins was asked if he was hopeful about the future of bird conservation. Yes, he answered, "all the time I see rays of sunshine. Sometimes they're way off in the distance, but . . . there's always hope that we can find more things to do to save more of the places we're interested in."

Robbins died in the spring of 2017, at the age of ninety-eight. Although he'd officially retired in 2005, he'd long continued to come into his office at Patuxent several days a week and kept giving presentations to bird clubs and going out birding himself until shortly before his death. A remembrance that ran in *The Washington Post* after his passing described him as "the Bruce Springsteen of birding." Everyone who knew him described him using words like "kind," "generous," and "humble." If only I'd written this book ten years earlier, I might have met and interviewed him myself. But his legacy lives on every June, when thousands of enthusiastic volunteers across the continent head out before dawn to count birds in the name of science, just as they have each year since 1966.

The Rise of eBird

Even if you don't want to get up at four in the morning and give up an entire June weekend, and even if your bird ID skills (like my own) are not quite expert level, you can still make a meaningful contribution to our understanding of birds' distribution and movements, thanks to an app called eBird.

The day after Christmas in December 2020, I was taking a chilly walk around my neighborhood when I heard a series of jeering bird-calls overhead. When I looked up, I saw a flock of five blue jays alighting in a tree above me. Where I live in southeastern Washington, blue jays are an unusual sight, and I pulled my phone out of my pocket to log the observation. Opening the app, I quickly noted my five jays and then tapped out the additional required details:

Date and Time	December 26, 2020, 12:46 p.m.
Location	Howard Street, Walla Walla, Washington
Observation Type	Incidental
Number of Observers	1

An "incidental" sighting indicated that this was not a complete list of all the birds I'd seen on my walk. Knowing that blue jays were out of the norm for the area, I added notes about the field marks I'd seen that clinched the ID—crests atop their heads, black "necklaces" across their chests, white markings in their tails.

In a way, eBird is the twenty-first-century answer to the BBS. But instead of relying on carefully screened volunteers who follow the same controlled procedure year after year, it lets anyone with an internet connection submit their observations of the birds around them, anytime, from anywhere in the world.

The idea to use the internet to collect data from bird-watchers originated with a collaboration between the Audubon Society and the Cornell Lab of Ornithology in the mid-1990s. In 1997, the two organizations launched BirdSource, a "World Wide Web site" through which a group of five hundred test users could submit their observations of birds at their feeders. The following year, they used the platform to host what they dubbed the BirdSource Great '98 Backyard Bird Count, encouraging "anyone with an interest in birds and access to the Internet" to submit counts of the birds they observed around their houses during a specific three-day period in February.

"The university's pipeline to the internet was so small at the time that we literally shut things down with the volume of checklists that were coming in," according to Steve Kelling, who had been hired to help run the project. "We only collected about fourteen thousand checklists, but that was fourteen thousand more than we ever expected we'd get."

Although an avid birder, Kelling was not a trained ornithologist. In fact, as he explained to me when I interrupted his retirement for a phone call one Wednesday morning in February, he originally came to Cornell as a lecturer in the biophysics department. But he was also interested in both computer science and birds and was involved with BirdSource and its successors from the very beginning, eventually becoming the Cornell Lab of Ornithology's director of information science.

The annual Great Backyard Bird Count, which Kelling helped launch, continues to this day. "But that foundation of that first year got us thinking, well, we have to build this national project," said Kelling, "and release it so people can submit their bird observations from anywhere, at any time."

The first question, of course, was exactly what data Kelling and his colleagues wanted to collect from bird-watchers. They settled on asking for six core pieces of information: who, where, when, what species, how many, and effort. (For the "where" part, they came up with a way for birders to click their location on a map to enter it into the database and ended up patenting the system they developed to do this, three years before the launch of Google Maps.) These are still the basic fields you'll be required to fill out if you log in to eBird to submit a checklist today: the date and location of your birding outing and how many of you there were; what bird species you saw and how many of each; and information on just how much effort you put into finding those birds, including how far you walked, how long you were out for, and, crucially, whether you're submitting a "complete checklist."

Submitting a complete checklist means that you made your best effort to identify and count every single bird around you. (My blue jay observation was not a complete checklist but an "incidental sighting," since I didn't keep a record of the other birds I saw on my walk.) Such checklists are at the core of what makes eBird data so useful, because often knowing what birds are *not* around is just as important as knowing what birds are. Without this information, scientists may not be able to figure out if a species they're studying is actually absent from a particular area, or if birders there just aren't bothering to report it because they don't find it interesting.

Funded by a grant from the National Science Foundation, eBird eventually launched in 2002. It was not an immediate success. "We were just getting a trickle of data. We kept working on trying to make it easier for people that submit their observations and, nothing. Nothing was happening, the birding community wasn't getting involved at all, and we were collecting maybe ten or fifteen thousand checklists a year, very little," said Kelling. "So my boss at the time, John Fitzpatrick [the director of the Cornell Lab of Ornithology], called me in and said you've got to tear this all down and put it back together."

Kelling hired two renowned birding guides, Chris Wood and Brian Sullivan, and assigned them the mission of making eBird more enticing for birders. Under Wood and Sullivan's guidance, eBird rolled out new features, such as letting birders track their "life lists" using eBird (many birders keep detailed lists of all the species they've seen in their lives) and making it easier to explore data submitted from their area to find new places to look for birds. It worked. "When we made this transition, eBird started to grow exponentially," said Kelling. By the time he retired in 2021, they were collecting more checklists every day than they had in the entire first three years of the project.

Today Kelling spends most of his time birding and, of course, submitting his sightings to eBird, the wildly popular platform that he helped develop. I asked him if that's ever a bit surreal, especially when he's surrounded by other birders also logging their observations on eBird with no idea that its creator is in their midst. "I've been very quiet about that. I always stayed behind the curtain," he said. But still, it must be pretty cool, right? "Yeah, yeah," he admitted. "I'm modestly very proud."

It took ten years from eBird's launch in 2002 for the platform to amass its first 100 million submitted observations. Its second 100 million took only two years. Today, more than 700,000 birders from every country in the world are uploading their bird checklists and contributing to eBird's ever-growing database. Though it originally only accepted data from the United States and Canada, eBird quickly expanded to include Mexico and then Central and South America before eventually going global in 2010. The year 2015 saw the launch of the official eBird mobile app, replacing a third-party app that many bird-watchers had already been using, letting eBirders track and submit their checklists from the field (as long as they have cell service).

The heart of eBird is its reviewers, a corps of extra-dedicated volunteers who spend their free time not only submitting their own observations but vetting and confirming the bird sightings of others in their regions. When a user submits a particularly unusual sighting—in February 2022, for example, birders on Twitter were sharing a screenshot of an eBird checklist from someone in Maryland who claimed to have had a herald petrel, a seabird from the South Pacific, at their backyard feeder—the system automatically flags it for a reviewer in their region to investigate. The reviewer typically reaches out to ask if the birder can provide more details or even a photo of the bird in question, working with them to help them understand how to document what they see to make their data as useful as possible. Ultimately, if a reviewer decides that a sighting is too implausible to be trusted, they can remove it from the database. One reviewer I spoke to in Washington state told me he typically spends four to five hours a week on this sort of thing, and that although he occasionally has to deal with grumps who get offended at the idea of someone questioning their bird identification skills, the vast majority of eBird users he interacts with are gracious and eager for his help. (To my surprise, he said he remembered my sighting of five blue jays. It was a big year for blue jays in the area, so it was a credible enough observation that he didn't contact me to inquire about it.)

Data from eBird has already found many applications for improving our knowledge of migration. The eBird database, for example, has helped make it clear that it's much more common than ornithologists once realized for migrating songbirds to follow different routes in spring and fall. Radar ornithologists like Kyle Horton, whom we met in chapter 3, can mine eBird data for clues about the specific species that might make up the massive movements of migrating birds showing up on weather radar. In combination with observations submitted to a similar Brazil-based website, eBird data even helped show for the first time that the muppetish, googly-eyed common potoo even deserves the title "migratory bird"; ornithologists had assumed that the reason no one heard potoos calling in parts of their range during winter was that they simply went silent during the nonbreeding season, but careful analysis of community science data showed that they're probably migrating within South America instead.

During the early months of the COVID-19 pandemic, eBird made it possible to study how the behavior of both birds and bird-watchers was changing due to lockdowns. Less traffic meant less noise and pollution, and eBird checklists showed that species ranging from bald eagles to hummingbirds were venturing into urban and developed areas more often. The closures of many parks and other natural areas, meanwhile, meant that birders were spending more time close to home, combing urban areas to find all those birds enjoying the unfamiliar quiet.

The most astonishing thing to come out of eBird in recent years, however, is what Cornell Lab of Ornithology scientists have dubbed eBird Status and Trends. Launched in 2018 and available on the eBird website, Status and Trends divides the globe up into a three-kilometer grid and provides estimates of how likely you are to find a given bird species at any spot on the globe, during any week of the year. On the eBird website, this takes the form of beautiful animated maps, with colors expanding and contracting across the globe to show a bird species' movements over the course of a year.

I called Tom Auer, a staffer at eBird since 2015, to learn more about how it works. An avid birder, Auer has a master's degree in geography and specializes in high-tech cartography. "I've always wanted to blend birds and mapping, ever since high school, so this is kind of a dream job," he said.

Data from eBird is, of course, inherently messy. "There's a lot of variability in how people go out in the world and encounter birds," said Auer. "It varies across countries, across individuals, and to understand how [bird] populations are changing from that information, we need to account for that variation in how people detect birds."

To do that, Auer and his colleagues use sophisticated computer models. "We break the world into small pieces of space and time, and then we run a machine learning model with eBird data and [land cover] data from within that region, produce predictions of how many birds there were if you standardized all that variation in observers and land cover, and then do that hundreds of times, because each individual model can be a little bit noisy." It would take an average laptop running continuously for 342 years to do this for a single bird species, and Status and Trends currently covers more than a thousand species. Luckily, the eBird team has access to a National Science Foundation supercomputer.

One of the first projects outside Cornell to use these models was headed up by Audubon Society researchers and published in 2021. California's Central Valley and the Colorado River delta have long been recognized as important regions for waterbirds, but ornithologists had never been able to confirm their suspicion that they were also crucial stopovers for migrating songbirds. Bill DeLuca—the same researcher who used geolocators to confirm blackpoll warblers' incredible open-ocean migration and who is now a migration ecologist with Audubon—wanted to change that and realized eBird might hold the answer.

Songbirds "are harder to survey" than waterfowl, DeLuca told me. "They're smaller, there's no one hunting them, and so it's been a difficult thing to estimate for a long time." But the Audubon Society has been working on bird conservation in the region for a long time, so demonstrating its importance for songbirds was an ideal project to tackle when he joined their Migratory Bird Initiative.

Using Status and Trends data as well as population estimates from other sources, DeLuca and his colleagues added up the total number of songbirds likely to be present each week in the areas they were interested in, then calculated roughly what percentage of the total population of each species that number was likely to represent. "The math behind it all is really easy," he said. "It's literally just adding and multiplying and getting proportions." (Although this particular research project was all computers and no field work, I was amused to note when I reached DeLuca via video call to talk about it that he was wearing a cap with a "New Hampshire mountains" graphic on it. The blackpoll warblers still have a hold on him, it appears.)

Their results showed that up to sixty-five million land birds migrate through California's Central Valley every spring, and a slightly smaller but still enormous number in the fall. For some species, this represented a significant percentage of their total North American population; for example, 59 percent of the world's tree swallows, 26 percent of the world's black-throated gray warblers, and 14 percent of the world's marsh wrens passed through these relatively small areas. DeLuca and his colleagues had confirmed the existence of a migratory bottleneck.

"Thinking about it in those terms simultaneously both blew our minds and scared the crap out of us," said DeLuca. "Entire populations, essentially, are dependent upon these really small areas at these really discrete times of year. And if just one piece of that whole puzzle gets removed, God only knows the cascading consequences that might have for some of those species." He hopes that having this concrete data to back up what ornithologists' guts were already telling them about the region will help bolster future conservation and restoration efforts.

In truth, this is all still just the "Status" part of Status and Trends. Set to launch in 2022, the trend modeling Auer and his colleagues are working on will combine eBird data from the past thirty years with land cover data to look at where bird populations are increasing or decreasing, and why, in incredible detail. For example, explained Auer, ornithologists will be able to look at how the amount of exotic grass cover in a specific habitat area has been changing and connect that directly to changes in local bird populations that result—like (to make up something hypothetical) 10 percent more cheatgrass in a specific Oregon county equals 2 percent fewer sage grouse each year.

The potential applications for bird conservation are immense. "There's a lot of opportunity to keep engaging communities, to inspire them to work with us to fill in gaps and keep monitoring places," said Auer. "I think this exciting new phase is going to include a little bit of science synergy, where we work with communities to help solve some of these problems together."

A Speck on the Globe

All this heady talk of big data and science synergy was still swirling around in my head the day I went birding with Bruce Toews. Walking along a gravel path on a blustery afternoon, we felt very far removed from supercomputers and cutting-edge machine learning, but this, after all, is where the data comes from.

Toews asked me if I was warm enough. "I have an extra hat if you need it," he said.

On one side of the path was a grassy field interspersed with shrubs. On the other, a line of low trees separated us from a fallow wheat field. I was keeping an eye out for northern shrikes or rough-legged hawks, birds that breed in the high Arctic but descend into the contiguous forty-eight states during the winter. All I saw was flock after flock of robins foraging in the trees. Still, as Emily Williams had made sure I knew, robins held some migration mysteries of their own.

The wind really was cold. The next time Toews offered me his spare hat, I said yes.

Although eBird is not a social networking site, users can create profile pages for themselves, and it's simple to log on to the website and pull up the profiles of the top eBirders in your county. The most prolific eBird user in my county by far was a well-known local naturalist who had worked for the U.S. Forest Service and now narrated nature documentaries for a regional TV station. The user with the second-most complete checklists for Walla Walla, however, was Toews, and he'd included his email address in his eBird profile. So here we were.

As we walked, he explained that he'd been interested in bird-watching ever since he was a kid, when his parents had introduced him to the hobby. It was in 2020, however, that his dedication to birds took hold with a new vigor. "I just realized, you know, I'm not getting any younger. I've been here [in Walla Walla] twenty-eight years and it's gone by fast, and in another twenty-eight years I'll be a very, very old man, if I live that long," he said. "I decided that I should stop working sixty or seventy hours a week and do more birding." Besides, with everything shut down due to the pandemic, how else was he going to spend his free time?

He declared 2020 his "big year," using eBird to track the birds he saw on his outings, and managed to find 239 different bird species in Walla Walla County—more than he was expecting. "And then I kind of got addicted to it, so last year was another big year, and this year is another big year, every year is a big year!" He hoped 2022 would be the year he broke 250.

Reaching a viewpoint from which we could look down at the reservoir, we stopped to count the waterfowl below us. "Two common mergansers—four—six . . ." We also tallied four great blue herons and a ring-necked duck. I noticed he didn't pull out a smartphone to put the totals into the eBird app, but Toews explained that he usually just leaves an audio recorder running and counts out loud as he birds, then enters the numbers in later. No recording today, though. "I'll just remember. This isn't that many birds."

Toews has noticed changes in the local bird community since he moved here a quarter century ago. There don't seem to be as many Lewis's woodpeckers around as there used to be; scrub jays, which used to be unheard of in the region, wander closer and closer to Walla Walla each year. He likes the idea that the data he collects could help unravel what's behind these range shifts, be it climate change or other factors. "I'm not a scientist by any means," he said, "but I have fun counting things, and I like that it makes a difference."

As we turned to hike back toward where we'd left our cars, the conversation turned philosophical. "I think that humans get so myopic, all they look at is their phone screens and TV screens, and they never get out, and they lose the sense of wonder. That's not living, you know?" he said. "Of course I can't criticize, everyone has their own thing, but I would probably be in a mental institution if I just sat in front of a screen all week and never got out birding."

A few months before our walk, Toews had undergone surgery for colon cancer. The prognosis was good, he told me, but it had made him even more aware of how uncertain our lives really are. "It's just great to get out and realize there's a bigger world than my little problem, I'm just a speck on the globe here. Life will go on without me. But at the same time, you know, it's very wholesome to have a diversion that helps you realize that there's still so much you want to do, you want to see."

Suddenly I noticed that the wind was gone. The setting sun had finally dipped down below the clouds, bathing the fields and shrubs in golden light. The robins were still flitting from one small tree to the next, their breasts glowing crimson.

"It gives you a little jolt of fighting spirit." Toews paused. "It makes you realize, I want to do some more birding."

Conclusion
Sky Full of Hope

Bruce Toews is right. When you're facing a scary, life-altering crisis, nothing resets your thinking quite like a morning spent outside with the birds.

In the spring of 2020, while the world was shutting down for the initial round of COVID-19 lockdowns, I was getting a cancer diagnosis of my own and preparing to begin chemotherapy. When the year started, I had no way of foreseeing the twin catastrophes it would bring, but I had made a New Year's resolution that would ultimately be critical for maintaining my mental health through that difficult year: I'd decided to keep a bird "year list" of my own, keeping track of how many bird species I observed over the course of 2020, something I hadn't done in a long time.

I've written about my strange, bittersweet cancer/pandemic year list before in other places, but what I remember most about that spring and summer are the flashes of color that lit up an otherwise dark time. I had left my job at the American Ornithological Society, I'd pulled my son out of day care, my world had contracted in a million tiny ways, but each Saturday my husband, toddler, and I would head out for a bird walk at the little nature area near our house. If I'd had a chemo treatment the day before, it would generally be a *short* walk, but there was always something new to see and hear: the bright turquoise plumage of returning lazuli buntings, the "sweet, sweet, a little more sweet" of the yellow warblers that suddenly filled the shrubs, male black-chinned and calliope hummingbirds buzzing around each other as they competed for territories.

By the end of that summer, the doctors had declared me cancer-free, and I'd begun working on the proposal that ultimately became the book you hold in your hands. Birding kept me going in part because, as Toews described, it provided a reminder that there was still a whole world out there beyond my own fears and troubles. Birding gave me hope.

But with the constant bombardment of bad news about climate change, plastic pollution, habitat loss, and other environmental problems, is there hope for the migratory birds themselves? I've occasionally wondered, while writing this book, if all of these new technologies allowing us to understand the intricacies of migration with greater clarity than ever before are actually just helping us document, in excruciating detail, the final decline of one of the natural world's most inspiring phenomena.

So I went looking for something or someone to convince me otherwise.

Choosing Optimism

Since Chan Robbins first sounded the alarm about a new, worrying drop in migratory songbird populations beginning in the late 1970s, things have not improved.

In 2019, a massive new study that drew data from a wide range of sources, including the Breeding Bird Survey and weather radar archives, reached a startling conclusion: North America is home to 2.9 billion fewer birds now than in 1970, a decline of around 29 percent. Land birds, waterbirds, and shorebirds; birds in forests and grasslands; birds that eat seeds and birds that eat insects—no matter how you categorize North America's feathered residents, the losses cross broad swaths of migratory species.

The "three billion birds" paper was splashed across the headlines when it came out, and a website launched to build on public interest in the study advised people on actions they could take to help birds, such as keeping cats (notorious bird killers) indoors and incorporating native plants into their lawns. But in the face of such monumental declines, it seems clear that individual efforts to make our homes and lives more bird-friendly will only go so far.

Are new tracking technologies the answer? We know more now than we ever have before about where migratory birds go, and when, and what threats they face along the way as a result, thanks to the techniques described in this book and the tireless efforts of the researchers who've developed, refined, and applied them. But it's surprisingly hard to find concrete examples of new information on the migration patterns of a specific species leading directly to better on-the-ground conservation.

Stories like that are out there, and the ones that do exist are striking. When I talked to the British ornithologist Nigel Clark about his efforts to track endangered spoon-billed sandpipers by satellite, he described officials in China quickly taking down nets and putting up warning signs at a newly identified stopover site. Earlier, in the 1990s, satellite transmitters let researchers follow mysteriously declining Swainson's hawks to their wintering grounds in South America. There, they found large numbers of them dying after exposure to the pesticide monocrotophos, and multiple countries soon made a plan to ban the use of the pesticide in areas important to the hawks. More recently, land was added to a nature preserve to protect a patch of northern Colombia that geolocators had shown was especially important to prothonotary warblers, and a coalition of countries agreed to work toward protecting a newly identified hot spot for seabirds in the northern Atlantic. But I'd expected to find dozens of these examples and instead came up with only a handful.

To better understand how technologies such as genoscapes and miniaturized tracking devices fit into the broader struggle to turn around declines in migratory bird populations—and, I hoped, be convinced that there was still cause to feel hopeful—I ended up calling two leaders in the field of migratory bird conservation. Pete Marra, a co-author of the three billion birds paper who used to be the head of the Smithsonian Migratory Bird Center and is now a professor at Georgetown University, and Jill Deppe, senior director of the Audubon Society's Migratory Bird Initiative, both agreed to chat with me. (Earlier in her career, Deppe worked on the pre-Motus automated radio telemetry system on the Yucatán Peninsula and traveled to Bill Cochran's home in Illinois so he could train her on how to use the antennas in his basement workshop. Ornithology truly is a small world.)

"How would you describe the current state of migratory bird conservation?" is such a broad question that I felt sheepish even asking it, but Marra and Deppe were both game. "We are in the sixth mass extinction," Marra told me bluntly. "It is the process of extinction that we're watching. It's just a question of what we do at this point. And so we all, as ornithologists, as birders, as people who care for our planet, our common home, can act now to try to do something, and that involves using the science to understand why these species are declining."

It would be easy to think that with the many, many studies published in recent decades that examine bird migration in ever-greater detail, the time for data collection is over and we should be devoting all of our resources to preserving habitat and reducing threats to migrants. But Marra and Deppe agreed that, surprisingly often, we *still* don't know exactly what threats are behind the declines of specific species. "I think right now, we're missing a lot of information on migration. We know the birds are declining, but we're still missing a lot of information on that migration period and how things are connected," said Deppe. "More and more people are interested in [migratory bird conservation]; more and more people who aren't scientists are starting to understand. So we're at a really pivotal moment where we understand the problem, and we're starting to have some tools at our disposal to figure out what the solutions are."

According to Marra, for example, until very recently no one knew why the western population of yellow-billed cuckoos seemed to be crashing, and there was no detailed information on where it went on migration. He and the postdoc Calandra Stanley placed tracking devices on cuckoos for the first time and found that the western birds were traveling to a region of South America being decimated by soybean agriculture.

Mystery solved—but not necessarily problem solved. There is an urgent need to consolidate data currently scattered across dozens of subscription-only scientific journals and get it to the conservationists and policy makers who need it, and to involve local communities (including indigenous communities) in seeking solutions to environmental problems that work for both birds and people. It isn't as simple as "this is important bird habitat, so you can't farm here anymore." "If you're going to invest millions of dollars in a conservative area, you want to be sure that it's really important. So you won't do it based on one study, you need to think about the bigger picture," said Deppe. "And any conservation solution is going to need to be one that includes people, benefits for people *and* birds." Ultimately, after all, "birds and people, we need the same things."

Marra is one of the co-founders of the Road to Recovery initiative, a project that grew out of the three billion birds paper as an attempt to "stop the bleeding" by identifying more of these apparent "smoking guns" for the fastest-declining species in North America and figuring out how to act on them. He and his colleagues have identified twenty-two "very high urgency" bird species and seventeen additional "high urgency" species, plus thirty-two they've labeled as "data deficient," meaning we don't even have a good grasp of just how much trouble they're in. Both the Road to Recovery and Audubon's Migratory Bird Initiative—which focuses more on identifying geographic areas of high importance to broad groups of species, such as the land bird migration bottleneck in California's Central Valley that they identified with eBird data—are keen to work across international borders and incorporate "social" science, aimed at improving humans' and birds' ability to coexist.

The path forward is not an easy one. But Marra and Deppe both insisted that they're hopeful about the future.

"I'm very optimistic," said Marra. "We've done it in the past, we've corrected [other environmental] issues, and we now have to deal with climate change. So while it seems like there's this overwhelming deluge of negative issues, and challenges with getting people on board with these things, and constant pressures on nature, I choose to be optimistic and hopeful just because I don't see any benefit of being pessimistic or having a lack of hope. I just don't choose to take that route.

"Don't get me wrong," he added, "there are times when I might be negative, or spiraling down into a pit of agony. But I'm not gonna do that."

Swan Song

When I'm spiraling down into a pit of agony of my own, I go birding.

The week after I went birding with Bruce Toews, I returned to the lake by myself on a quiet Friday morning. I couldn't see the water from the parking lot, but when I got out of my car, I heard a loud racket of unfamiliar birdcalls coming from that direction. Not Canada geese, I thought—maybe snow geese? (My birding-by-ear skills are still, shall we say, in development.) I immediately started working my way along a trail to get a view of whatever birds were making all the noise.

The morning sun was in my eyes, but I could make out a flock of perhaps fifty large white birds on the water. Snow geese it was, then—a fairly common species here in winter and during migration. But then their calls reached a crescendo, and all at once they lifted off, coming straight toward me, the beating of their wings adding to the din. As they approached, I got a better look and realized what I was seeing. Not geese. Swans.

The flock of tundra swans wheeled as one until they were pointed north, then disappeared over a rise. In May those birds would be raising babies on the shores of the Arctic Ocean.

I don't know what the next year, or the next decade, or the next century holds for their descendants—or for mine. But I do know that all over the world, in Colombia and Mongolia and the Netherlands and here in my own community, scientists, bird-watchers, and activists of all sorts are dedicating themselves to making sure that migratory birds will continue to be here, continue to knit the globe together with their journeys, and continue to bring hope to all who see them.

Acknowledgments

A lot of writing a book consists in sitting by yourself and staring at a computer screen, but I've learned in the past two years that it is far from a solitary endeavor.

My first thank-you goes to the amazing doctors and (especially!) nurses at the Providence St. Mary Regional Cancer Center, without whom I probably wouldn't have been around to write this book. Some bits of the research and writing that eventually became the proposal that became this book were literally done while I was sitting in the chemo infusion chair. Luckily, the infusion room has free Wi-Fi.

Thank you to Michael Metivier, who first told me he thought this sounded like a solid book idea and encouraged me to develop it, and to Ben Goldfarb, Jennifer Howard, and the other writers on Twitter who answered questions and shared tips with this first-time author. Thank you to my agent, Michelle Tessler, and to Maddie Pillari and Lisa Sharkey at HarperCollins, for believing in this book and guiding it along.

Thanks to each and every one of the scientists, birders, and others quoted in this book, plus everyone else whose names and stories *didn't* necessarily make it into the book but who answered questions, sent me papers and photos, and just generally cheered me on. This includes Steve Albert, Paige Bailey, Liv Baker, Dan Baldassarre, Matt Boone,

Alice Boyle, Eli Bridge, Jeff Buler, Matt Carling, Jamie Chambers, Kristen Covino, Jenna Curtis, Mike Dardeau, Glen Fowler, Dominic Garcia-Hall, Kate Goodenough, Autumn-Lynn Harrison, Julie Heath, David Holmes, Wesley Honeycutt, Marshall Iliff, Anna Catalina Salsac Jimenez, Lance Jordan, Jeff Kelly, Alex Lees, Dan Mennill, Harold Mills, Tim O'Connell, Keith Pardieck, Ellen Paul, Molly Paul, Richard Phillips, Steve Portugal, Rob Robinson, Clark Rushing, Lisa Schibley, Joe Siegrist, Tom Strikwerda, Lynn Sweet, Heidi Trudell, Bill Tweit, Joe Wall, Sean Walla, Julián García Walther, Mike Ward, and of course the wonderful crews of field assistants in Montana and Illinois who helped ensure I got to see the science I was writing about in action. If you helped me or encouraged me in any way while I was writing this book and I have left your name off this list, I apologize. It meant more than you know.

Thank you to Erin Hvizdak and David Luftig, Washington State University librarian superheroes who made repeated forays into the stacks to scan old *Audubon Magazine* articles for me despite my having no official affiliation with WSU; the Walla Walla Public Library librarians who located a copy of an obscure book on wildlife tracking at a library in Alaska and had it lent across state lines for me; Ginny Roth at the U.S. National Library of Medicine and Susan Mae Braxton and Cathy Bialeschki at the Illinois Natural History Survey for helping me track down old photos and make sure I was crediting them correctly; and the staff who keep the archives at SORA, JSTOR, and the Biodiversity Heritage Library up and running. Thank you also to the crew at Colville Street Patisserie for keeping me caffeinated and fueled with pastries and generally cheering me on. This is not the first book by a Walla Walla author that mentions the Patisserie in the acknowledgments!

Thank you to Sarah Swanson and Scarlett Rebman for reading and commenting on early drafts, Nancy Steinberg for being the best book accountability buddy anyone could want, and my parents for helping out with child care during COVID craziness so I could keep working. And most of all, thank you to Evan, for everything, always.

Finally, I want to acknowledge that all of the research and writing for this book was done on stolen land. My little home office sits on the occupied land of the Umatilla, Walla Walla, and Cayuse peoples, and my research trips took me to the stolen homes of the Blackfoot, Kaskaskia, Ktunaxa, Choctaw, and Shoshone-Bannock peoples, among (I'm sure) others.

Like other branches of science from which minorities have historically been excluded, the field of ornithology is still disproportionately dominated by white scientists, a fact reflected in the pages of this book. We all have more work to do, and I will be donating a portion of my proceeds from this book to an organization that provides scholarships to BIPOC (Black, indigenous, and/or person of color) birders interested in STEM fields.

Further Reading and Resources

Audubon Christmas Bird Count (www.audubon.org/conservation /science/christmas-bird-count). How to participate in the world's oldest continuously running community science program.

BirdCast (birdcast.info). Radar-based forecasts of bird migration intensity across the United States.

Bird Genoscape Project (www.birdgenoscape.org). More about the Bird Genoscape Project's research and findings.

eBird (ebird.org). How to start uploading your own sightings to eBird, plus lots of ways to explore the data that's already been accumulated.

Icarus (www.icarus.mpg.de/en). Information on the new Icarus wildlife tracking system based on the International Space Station.

Motus Wildlife Tracking System (motus.org). Information on the Motus automated radio telemetry system.

Movebank (www.movebank.org). A website where anyone can access maps of the movements of animals currently being tracked around the world.

North American Breeding Bird Survey (www.pwrc.usgs.gov/bbs). In-formation on the long-running annual survey founded by Chan Robbins, including how to volunteer.

Old Bird (www.oldbird.org). Information and resources related to nocturnal flight call recording, including gear to set up your own recording station.

Terra Project (www.terralistens.com). A planned device for commu-nity scientists to detect nocturnal flight calls and Motus-tagged birds passing over their yards.

USGS Bird Banding Laboratory (www.usgs.gov/labs/bird-banding -laboratory). Information on bird banding in the United States, including how to report a sighting of a banded bird.

Notes

Introduction | Where Do the Birds Go?

2 an old story: "Raven and Goose-Wife," Alaska Native Knowledge Network, accessed March 7, 2022, www.ankn.uaf.edu/NPE/Cultural Atlases/Yupiaq/Marshall/raven/RavenandGooseWife.html.

2 crevices in trees: Alexander Lee, "The Great Migration Mystery," *History Today*, May 5, 2020, www.historytoday.com/archive/natural -histories/great-migration-mystery.

2 transmogrified into European robins: Matt Simon, "Fantastically Wrong: The Scientist Who Thought That Birds Migrate to the Moon," *Wired*, Oct. 22, 2014, www.wired.com/2014/10/fantastically-wrong -scientist-thought-birds-migrate-moon/.

2 Olaus Magnus: John Lienhard, "Ancient Explanations of Bird Migration," *Engines of Our Ingenuity*, June 8, 2007, uh.edu/engines/epi22 28.htm.

3 flying to the moon: Simon, "Fantastically Wrong."

3 an unfortunate stork: Nancy J. Jacobs, "Africa, Europe, and the Birds Between Them," in *Eco-cultural Networks and the British Empire* (London: Bloomsbury Academic, 2015).

3 Migration is simply: "The Basics of Bird Migration: How, Why, and Where," All About Birds, Aug. 1, 2021, www.allaboutbirds.org/news /the-basics-how-why-and-where-of-bird-migration/.

4 two competing theories: "The Evolution of Bird Migration," All

About Birds, April 11, 2017, www.allaboutbirds.org/news/the
-evolution-of-bird-migration/.

4   navigation techniques: "Basics of Bird Migration."

5   quantum physics: Peter J. Hore and Mouritsen Henrik, "The Quantum Nature of Bird Migration," *Scientific American*, April 2022, doi
.org/10.1038/scientificamerican0422-26.

5   Visits to websites: Lori Mack, "Bird-Watching Soars During Pandemic," Connecticut Public Radio, May 7, 2020, www.ctpublic.org
/mental-health/2020-05-07/bird-watching-soars-during-pandemic.

5   Retailers couldn't keep up: Neel Dhanesha, "Birdwatching Is a
Bright Spot in a Pandemic-Stricken Economy," Audubon, Aug. 6, 2020,
www.audubon.org/news/birdwatching-bright-spot-pandemic
-stricken-economy.

5   Facebook groups: Elaine Glusac, "Birdwatching Becomes More
Popular During the Pandemic," AARP, June 16, 2020, www.aarp.org
/home-family/friends-family/info-2020/bird-watching-popularity
.html.

One | A Bird in the Hand

9   recorded fates: Annie Lindsay, email message to author, Dec. 4, 2020.

10   "When banded birds": "Why Do We Band Birds?," accessed
March 8, 2021, www.usgs.gov/centers/pwrc/science/why-do-we-band
-birds?qt-science_center_objects=0#qt-science_center_objects.

11   leg irritation: B. Calvo and R. W. Furness, "A Review of the Use and
the Effects of Marks and Devices on Birds," *Ringing & Migration* 13,
no. 3 (1992): 129–51, doi.org/10.1080/03078698.1992.9674036.

11   fewer than six: Erica N. Spotswood et al., "How Safe Is Mist Netting? Evaluating the Risk of Injury and Mortality to Birds," *Methods in
Ecology and Evolution* 3, no. 1 (2012): 29–38, doi.org/10.1111/j.2041
-210X.2011.00123.x.

12   According to Bird Banding Laboratory: Antonio Celis-Murillo, email message to author, Oct. 26, 2020.

12   a million bands: "USGS Celebrates 100 Years of Bird Banding Lab," accessed March 15, 2021, www.usgs.gov/news/usgs-celebrates-100 -years-bird-banding-lab.

12   "reward bands": "Reward Bands," accessed March 15, 2021, www .fws.gov/birds/surveys-and-data/bird-banding/reward-bands.php.

12   behavior of house finches: Gilbert Cant and Putman Geis, "The House Finch: A New East Coast Migrant?," *Eastern Bird Banding Association News* 24, no. 4 (1961): 102–7.

12   And they didn't stop there: K. P. Able and J. R. Belthoff, "Rapid 'Evolution' of Migratory Behaviour in the Introduced House Finch of Eastern North America," *Proceedings of the Royal Society B: Biological Sciences* 265, no. 1410 (1998): 2063–71, doi.org/10.1098/rspb.1998 .0541.

13   something strange and wonderful: Ian C. T. Nisbet, "Autumn Migration of the Blackpoll Warbler: Evidence for Long Flight Provided by Regional Survey," *Bird-Banding* 41, no. 3 (1970): 207–40, doi.org /10.2307/4511673.

14   "These they invariably removed": John James Audubon, *Ornithological Biography*, 5 vols. (Project Gutenberg eBook, n.d.), vol. 2, www .gutenberg.org/files/57191/57191-h/57191-h.htm.

14   "favorable to his legacy": Matthew R. Halley, "Audubon's Famous Banding Experiment: Fact or Fiction?," *Archives of Natural History* 45, no. 1 (2018): 118–21, doi.org/10.3366/anh.2018.0487.

14   the breasts of snow buntings: Ernest Thompson Seton, "Early Bird Banding," *Auk* 38, no. 4 (1921): 611, sora.unm.edu/node/12431.

15   Boy Scouts of America: "Ernest Thompson Seton," Order of the Arrow, BSA, accessed March 15, 2021, oa-bsa.org/history/ernest -thompson-seton.

15   Hans Christian Cornelius Mortensen: Harold B. Wood, "The History of Bird Banding," *Auk* 62, no. 2 (1945): 256–65.

15   Cole was inspired: Leon J. Cole, "Suggestions for a Method of Studying the Migrations of Birds," *Report of the Michigan Academy of Science* 3 (1901): 67–70, www.biodiversitylibrary.org/item /26681.

16   a Smithsonian scientist: Wood, "History of Bird Banding."

16   One of Taverner's bands: P. A. Taverner, "Tagging Migrants," *Auk* 23, no. 2 (1906): 232.

16   Cole had grander ambitions: Leon J. Cole, "The Tagging of Wild Birds as a Means of Studying Their Movements," *Auk* 26, no. 2 (1909): 137–43.

16   "disappointingly small": Leon J. Cole, "The Early History of Bird Banding in America," *Wilson Bulletin* 34, no. 2 (1922): 108–15.

17   Cole reported back: Leon J. Cole, "The Tagging of Wild Birds: Report of Progress in 1909," *Auk* 27, no. 2 (1910): 153–68.

17   "Gentlem dear sirs": Cole, "Early History of Bird Banding in America."

18   professional and amateur naturalists: "About the Society," Linnaean Society of New York, accessed March 15, 2021, www.linnaeannewyork .org/about-the-society/.

18   Bureau of Biological Survey: P. A. Taverner, "Bird Banding Work Being Taken Over by the United States Bureau of Biological Survey," *Canadian Field-Naturalist* 34 (1920): 158–59, www.biodiversitylibrary.org /item/17534.

18   forerunner of the U.S. Fish: "United States Bureau of Biological Survey," Smithsonian Institution Archives, siarchives.si.edu/collections /auth_org_fbr_eaco13.

19   "permanent segregation": Leon J. Cole, "Biological Eugenics: Relation of Philanthropy and Medicine to Race Betterment—Study of

Genetics Shows That No Race Can Be Bred Immune to All Diseases or Defects—Nevertheless, Medicine and Charity Must Pay More Attention to Heredity," *Journal of Heredity* 5, no. 7 (1914): 305–12, doi .org/10.1093/oxfordjournals.jhered.a107879.

19   birth control: Gordon E. Dickerson and Arthur B. Chapman, "Leon Jacob Cole, 1877–1948: A Brief Biography," *Journal of Animal Science* 67, no. 7 (1989): 1653–56, doi.org/10.2527/jas1989.6771653x.

19   "exceptional and sustained contributions": "Fellows, Honorary Fellows & Elective Members," American Ornithological Society, accessed March 15, 2021, americanornithology.org/about/fellows-elective -members/.

19   "What happened, I never knew": Robert A. McCabe, "Wisconsin's Forgotten Ornithologist: Leon J. Cole," *Passenger Pigeon* 41, no. 3 (Fall 1979), digicoll.library.wisc.edu/cgi-bin/EcoNatRes/EcoNatRes-idx? type=article&did=EcoNatRes.pp41n03.RMcCabe&id=EcoNatRes .pp41n03&isize=M.

20   "Wild Goose Jack": Manly F. Miner, "Migration of Canada Geese from the Jack Miner Sanctuary and Banding Operations," *Wilson Bulletin* 43, no. 1 (1931): 29–34.

20   "Never of very rugged constitution": S. Charles Kendeigh, "In Memoriam: Samuel Prentiss Baldwin," *Auk* 57, no. 1 (1940): 1–12.

20   retired from practicing law: "Baldwin, Samuel Prentiss," *Encyclopedia of Cleveland History*, May 11, 2018, case.edu/ech/articles/b/baldwin -samuel-prentiss.

21   treatise on his trapping and banding efforts: S. Prentiss Baldwin, "Bird-Banding by Means of Systematic Trapping," *Abstract of the Proceedings of the Linnaean Society of New York* 31 (1919): 23–56, www.bio diversitylibrary.org/item/49176.

22   stumbled across immigrants: Joseph Grinnell, "Bird Netting as a Method in Ornithology," *Auk* 42, no. 2 (1925): 245–51.

22   reports from the island nation: Hugh H. Genoways, Suzanne B. McLaren, and Robert M. Timm, "Innovations That Changed

Mammalogy: The Japanese Mist Net," *Journal of Mammalogy* 101, no. 3 (2020): 627–29, doi.org/10.1093/jmammal/gyaa055.

22   "The net had slipped": Walter W. Dalquest, "Netting Bats in Tropical Mexico," *Transactions of the Kansas Academy of Science* 57, no. 1 (1954): 1–10, doi.org/10.2307/3625633.

24   after a reorganization: P. A. Buckley et al., "The North American Bird Banding Program: Into the 21st Century," *Journal of Field Ornithology* 69, no. 4 (Autumn 1998): 511–29.

26   begins earlier and takes longer: Josh Van Buskirk, Robert S. Mulvihill, and Robert C. Leberman, "Variable Shifts in Spring and Autumn Migration Phenology in North American Songbirds Associated with Climate Change," *Global Change Biology* 15, no. 3 (2009): 760–71, doi .org/10.1111/j.1365-2486.2008.01751.x.

26   Banding data from Europe: Esa Lehikoinen, Tim H. Sparks, and Mecislovas Zalakevicius, "Arrival and Departure Dates," in *Advances in Ecological Research*, vol. 35, *Birds and Climate Change*, ed. Anders P. Møller, Wolfgang Fiedler, and P. Berthold (Amsterdam: Academic Press, 2004), 1–31, doi.org/10.1016/S0065-2504(04)35001-4.

27   "microevolution": Josh Van Buskirk, Robert S. Mulvihill, and Robert C. Leberman, "Phenotypic Plasticity Alone Cannot Explain Climate-Induced Change in Avian Migration Timing," *Ecology and Evolution* 2, no. 10 (2012): 2430–37, doi.org/10.1002/ece3.367.

27   one day later per decade: Sara R. Morris et al., "Fall Migratory Patterns of the Blackpoll Warbler at a Continental Scale," *Auk* 133, no. 1 (2016): 41–51, doi.org/10.1642/AUK-15-133.1.

27   half a day earlier: Kristen Covino et al., "Spring Migration of Blackpoll Warblers Across North America," *Avian Conservation and Ecology* 15, no. 1 (2020), doi.org/10.5751/ACE-01577-150117.

27   black-throated blue warbler: Kristen M. Covino, Kyle G. Horton, and Sara R. Morris, "Seasonally Specific Changes in Migration Phenology Across 50 Years in the Black-Throated Blue Warbler," *Auk* 137, no. 2 (2020), doi.org/10.1093/auk/ukz080.

27   that's a problem: "When Timing Is Everything: Migratory Bird Phenology in a Changing Climate," U.S. Geological Survey, Feb. 10, 2017, www.usgs.gov/center-news/when-timing-everything-migratory -bird-phenology-a-changing-climate.

Two | Looking and Listening

30   "The air seemed": O. G. Libby, "The Nocturnal Flight of Migrating Birds," *Auk* 16, no. 2 (1899): 140–46.

31   Ornithologists' best guess: Andrew Farnsworth, "Flight Calls and Their Value for Future Ornithological Studies and Conservation Research," *Auk* 122, no. 3 (2005): 733–46, doi.org/10.1093/auk/122.3.733.

31   Recent research shows: Zach G. Gayk, Richard K. Simpson, and Daniel J. Mennill, "The Evolution of Wood Warbler Flight Calls: Species with Similar Migrations Produce Acoustically Similar Calls," *Evolution* 75, no. 3 (2021): 719–30, doi.org/10.1111/evo.14167.

32   Sometimes they made educated guesses: Winsor M. Tyler, "The Call-Notes of Some Nocturnal Migrating Birds," *Auk* 33, no. 2 (1916): 132–41.

32   Stanley Ball: Stanley C. Ball, *Fall Migration on the Gaspé Peninsula* (New Haven, Conn.: Peabody Museum, 1952).

32   The two met: These details according to my interview with Bill Evans, March 31, 2021.

32   For their first attempt: Richard R. Graber and William W. Cochran, "An Audio Technique for the Study of Nocturnal Migration of Birds," *Wilson Bulletin* 71, no. 3 (1959): 220–36.

33   in 1960 they published: Richard R. Graber and William W. Cochran, "Evaluation of an Aural Record of Nocturnal Migration," *Wilson Bulletin* 72, no. 3 (1960): 253–73.

39   user-friendly software: Leonida Fusani and Renato Massa, "Software Review," *Ethology Ecology & Evolution* 6, no. 2 (1994): 249–52, doi .org/10.1080/08927014.1994.9522999.

41   over the Gulf of Mexico: Andrew Farnsworth and Robert W. Russell, "Monitoring Flight Calls of Migrating Birds from an Oil Platform in the Northern Gulf of Mexico," *Journal of Field Ornithology* 78, no. 3 (2007): 279–89, www.jstor.org/stable/40345964.

41   across Lake Erie: Claire E. Sanders and Daniel J. Mennill, "Acoustic Monitoring of Nocturnally Migrating Birds Accurately Assesses the Timing and Magnitude of Migration Through the Great Lakes," *Condor* 116, no. 3 (2014): 371–83, doi.org/10.1650/CONDOR-13-098.1.

41   coast of Rhode Island: Adam D. Smith, Peter W. C. Paton, and Scott R. McWilliams, "Using Nocturnal Flight Calls to Assess the Fall Migration of Warblers and Sparrows Along a Coastal Ecological Barrier," *PLoS ONE* 9, no. 3 (2014): e92218, doi.org/10.1371/journal.pone.00 92218.

41   seabirds called scoters: "Citizen Science Reveals Nocturnal Scoter Migration Routes," BirdGuides, April 6, 2020, www.birdguides.com /articles/migration/citizen-science-reveals-nocturnal-scoter -migration-routes/.

41   attracted to artificial lights: Matthew J. Watson, David R. Wilson, and Daniel J. Mennill, "Anthropogenic Light Is Associated with Increased Vocal Activity by Nocturnally Migrating Birds," *Condor* 118, no. 2 (2016): 338–44, doi.org/10.1650/CONDOR-15-136.1.

42   Martin Minařík: Martin Minařík, Facebook messages to author, April 5–6, 2021.

43   "My attention was at once": W. E. D. Scott, "Some Observations on the Migration of Birds," *Bulletin of the Nuttall Ornithological Club* 6, no. 2 (1881): 97–100, sora.unm.edu/sites/default/files/Nutt_vol6_n02_91 .pdf.

43   modern data shows: Gustave Axelson, "New BirdCast Analysis Shows How High Migrating Birds Fly," All About Birds, Oct. 13, 2021, www.allaboutbirds.org/news/new-birdcast-analysis-shows-how-high -migrating-birds-fly/.

44   In the early 1940s: George H. Lowery Jr., "Evidence of Trans-Gulf Migration," *Auk* 63, no. 2 (1946): 175–211.

44 LSU Museum of Zoology: Thomas R. Howell and John P. O'Neill, "In Memoriam: George H. Lowery Jr.," *Auk* 98, no. 1 (1981): 159–66.

44 quirky technique: Robert J. Newman, "Wings Across the Moon," *Audubon Magazine*, Aug. 1952.

44 bird-loving astronomer: Glen Cushman, "William A. Rense (1914– 2008)," *Bulletin of the AAS* 41, no. 4 (2009), baas.aas.org/pub/william -rense-1914-2008/release/1.

45 "That night, he set up": Miyoko Chu, *Songbird Journeys* (New York: Walker, 2006).

45 Lowery didn't give up easily: Lowery, "Evidence of Trans-Gulf Migration."

45 "Flight studies by means": Newman, "Wings Across the Moon."

46 graduate student in 1945: Frances C. James, "In Memoriam: Robert James Newman," *Auk* 106, no. 3 (1989): 464–65.

46 Lowery—or "Doc": Details in this and the following paragraph are from author's interviews and correspondence with Sidney Gauthreaux, Jan. 2021.

47 "must get up": Newman, "Wings Across the Moon."

47 "The Museum began": Robert Newman, "Hour-to-Hour Variation in the Volume of Nocturnal Migration in Autumn" (PhD diss., Louisiana State University, Baton Rouge, 1956), digitalcommons.lsu.edu /gradschool_disstheses/178.

48 twenty-five hundred volunteers: George H. Lowery Jr. and Robert J. Newman, "Direct Studies of Nocturnal Bird Migration," in *Recent Studies in Avian Biology*, ed. Albert Wolfson (Urbana: University of Illinois Press, 1955).

48 how-to pamphlet: George H. Lowery Jr. and Robert J. Newman, *Studying Bird Migration with a Telescope* (Baton Rouge: Louisiana State University Museum of Zoology, 1963).

49 their magnum opus: George H. Lowery Jr. and Robert J. Newman,

"A Continentwide View of Bird Migration on Four Nights in October," *Auk* 83, no. 4 (1966): 547–86.

50   dubbed LunAero: Wesley T. Honeycutt et al., "LunAero: Automated 'Smart' Hardware for Recording Video of Nocturnal Migration," *HardwareX*, April 1, 2020, e00106, doi.org/10.1016/j.ohx.2020.e00106.

50   phenomena like social behavior: Eli Bridge and Wesley Honeycutt, email messages to author, March 31, 2021.

Three | Chasing Angels

52   Irven Buss: Irven O. Buss, "Bird Detection by Radar," *Auk* 63, no. 3 (1946): 315–18.

52   sandpipers and pheasants: Richard E. Johnson, "In Memoriam: Irven O. Buss, 1908–1993," *Auk* 113, no. 3 (1996): 685.

54   "moving at between 5 and 80": Anthony D. Fox and Patrick D. L. Beasley, "David Lack and the Birth of Radar Ornithology," *Archives of Natural History* 37, no. 2 (2010): 325–32, doi.org/10.3366/anh.2010.0013.

54   RAF Fighter Command: David Hambling, "Weatherwatch: Planes, Rain, and Wartime Radar Angels," *Guardian*, March 27, 2021, www .theguardian.com/news/2021/mar/27/weatherwatch-planes-rain-and -wartime-radar-angels.

55   Darwin's famous finches: Ted R. Anderson, *The Life of David Lack: Father of Evolutionary Ecology* (Oxford: Oxford University Press, 2013).

55   join the war effort: "Obituaries," *Ibis* 115, no. 3 (1973): 421–41, doi .org/10.1111/j.1474-919X.1973.tb01982.x.

55   recounted his conversation: Ibid.

56   large white seabirds: David Lack, "Recent Swiss and British Work on Watching Migration by Radar," *Ibis* 100, no. 2 (1958): 286–87, doi .org/10.1111/j.1474-919X.1958.tb08806.x.

56   "At one meeting": Ibid.

56   1945 letter: David Lack and G. C. Varley, "Detection of Birds by Radar," *Nature* 156, no. 3963 (1945): 446, doi.org/10.1038/156446a0.

56   Other publications: H. A. C. McKay, "Detection of Birds by Radar," *Nature* 156, no. 3969 (1945): 629, doi.org/10.1038/156629a0.

57   switched gears entirely: Johnson, "In Memoriam."

58   first modern weather radar network: Roger C. Whiton et al., "History of Operational Use of Weather Radar by U.S. Weather Services. Part I: The Pre-NEXRAD Era," *Weather and Forecasting* 13, no. 2 (1998): 219–43, doi.org/10.1175/1520-0434(1998)013<0219:hoouow>2.0.CO;2.

59   Swarms of insects: Bruno Bruderer, "The Study of Bird Migration by Radar Part 1: The Technical Basis," *Naturwissenschaften* 84, no. 1 (1997): 1–8, doi.org/10.1007/s001140050338.

61   By comparing his radar data: Sidney A. Gauthreaux Jr., "Weather Radar Quantification of Bird Migration," *BioScience* 20, no. 1 (1970): 17–19, doi.org/10.2307/1294752.

62   Williams had continued to argue: George G. Williams, "Birds on the Gulf of Mexico," *Auk* 69, no. 4 (1952): 428–32.

62   sixty-five hundred feet or less: Emily B. Cohen et al., "How Do En Route Events Around the Gulf of Mexico Influence Migratory Landbird Populations?," *Condor* 119, no. 2 (May 2017): 327–43, doi.org/10.1650/CONDOR-17-20.1.

62   Ron Larkin: Ronald P. Larkin et al., "Radar Observations of Bird Migration over the Western North Atlantic Ocean," *Behavioral Ecology and Sociobiology* 4, no. 3 (1979): 225–64, doi.org/10.1007/BF00297646.

62   Timothy and Janet Williams: Timothy C. Williams et al., "Autumnal Bird Migration over the Western North Atlantic Ocean," *American Birds* 31, no. 3 (1977): 251–67.

63   "I watched an RV": Mike Dardeau (@mdardeau), Twitter, Sept. 15, 2020, 8:08 a.m., twitter.com/mdardeau/status/1305886291175800832.

65   rest and refuel: Jeffrey J. Buler and Deanna K. Dawson, "Radar

Analysis of Fall Bird Migration Stopover Sites in the Northeastern U.S.," *Condor* 116, no. 3 (2014): 357–70, doi.org/10.1650/CONDOR -13-162.1.

65   the Great Lakes: David V. Gesicki, Erica L. Cech, and Verner P. Bingman, "Detoured Flight Direction Responses Along the Southwest Coast of Lake Erie by Night-Migrating Birds," *Auk* 136, no. 3 (2019), doi.org/10.1093/auk/ukz018.

65   between spring and fall: Kyle G. Horton et al., "Seasonal Differences in Landbird Migration Strategies," *Auk* 133, no. 4 (2016): 761–69, doi.org/10.1642/AUK-16-105.1.

66   a billion birds: Scott R. Loss et al., "Bird–Building Collisions in the United States: Estimates of Annual Mortality and Species Vulnerability," *Condor* 116, no. 1 (2014): 8–23, doi.org/10.1650/CONDOR -13-090.1.

66   with funding from the EPA: "Developing and Implementing a Bird Migration Monitoring, Assessment, and Public Outreach Program for Your Community: The BirdCast Project" (Cincinnati: U.S. Environmental Protection Agency, 2001), nepis.epa.gov/Exe/ZyPURL .cgi?Dockey=30004I1W.txt.

69   available for download: "Partnering with Amazon Web Services on Big Data," NOAA National Centers for Environmental Information, accessed April 19, 2021, www.ncdc.noaa.gov/news/partnering-amazon -web-services-big-data.

70   A 2018 study: James D. McLaren et al., "Artificial Light at Night Confounds Broad-Scale Habitat Use by Migrating Birds," *Ecology Letters* 21, no. 3 (2018): 356–64, doi.org/10.1111/ele.12902.

70   another big-data project: Kyle G. Horton et al., "Bright Lights in the Big Cities: Migratory Birds' Exposure to Artificial Light," *Frontiers in Ecology and the Environment* 17, no. 4 (2019): 209–14, doi.org/10.1002 /fee.2029.

71   Campaigns in cities: "Existing Lights Out Programs," Audubon, March 16, 2015, www.audubon.org/conservation/existing-lights-out -programs.

71   American National Insurance: American National Insurance Com-
      pany, "American National Celebrates Earth Day by Committing to
      Protect Migratory Birds," GlobeNewswire News Room, April 29, 2019,
      www.globenewswire.com/news-release/2019/04/29/1811460/12861
      /en/American-National-Celebrates-Earth-Day-by-Committing-to
      -Protect-Migratory-Birds.html.

71   Laura Bush: "Thousands of Birds Will Die in Dallas Tonight Unless
      We Do One Simple Thing," *Dallas Morning News*, May 1, 2020, www
      .dallasnews.com/opinion/editorials/2020/05/01/thousands-of-birds
      -will-die-in-dallas-tonight-unless-we-do-one-simple-thing/.

71   In the fall of 2019: Kenneth V. Rosenberg et al., "Decline of the
      North American Avifauna," *Science* 366, no. 6461 (2019): 120–24, doi
      .org/10.1126/science.aaw1313.

Four | Follow That Beep

72   On the morning of May 13: William W. Cochran, "Orientation and
      Other Migratory Behaviours of a Swainson's Thrush Followed for 1500
      Km," *Animal Behaviour* 35, no. 3 (June 1987): 927–29, doi.org/10.1016
      /S0003-3472(87)80132-X.

72   the bird's skin: Arlo Raim, "A Radio Transmitter Attachment for
      Small Passerine Birds," *Bird Banding* 49, no. 4 (Autumn 1978): 326–32.

75   a small radar unit: Richard R. Graber, "Nocturnal Migration in
      Illinois—Different Points of View," *Wilson Bulletin* 80, no. 1 (1968): 36–71.

75   "exceptional facility": G. W. Swenson, "Reminiscence: At the Dawn
      of the Space Age," *IEEE Antennas and Propagation Magazine* 36, no. 2
      (1994): 32–35, doi.org/10.1109/74.275548.

75   the flow of electrons: Michael Riordan, "The Lost History of the
      Transistor," *IEEE Spectrum*, April 30, 2004, spectrum.ieee.org/tech
      -history/silicon-revolution/the-lost-history-of-the-transistor.

75   jet pilot's vital signs: Norman Lee Barr, "The Radio Transmission
      of Physiological Information," *Military Surgeon* 114, no. 2 (1954): 79–83,
      doi.org/10.1093/milmed/114.2.79.

75   cold-water suits: Etienne Benson, *Wired Wilderness* (Baltimore: Johns Hopkins University Press, 2010).

75   penguin egg: Carl R. Eklund and Frederick E. Charlton, "Measuring the Temperatures of Incubating Penguin Eggs," *American Scientist* 47, no. 1 (1959): 80–86, www.jstor.org/stable/27827248.

76   transmitters in woodchucks: Cobert D. LeMunyan et al., "Design of a Miniature Radio Transmitter for Use in Animal Studies," *Journal of Wildlife Management* 23, no. 1 (1959): 107–10, doi.org/10.2307/3797755.

76   "may help discover": Benson, *Wired Wilderness.*

77   very first data: Swenson, "Reminiscence."

79   small, simple, and compact: Benson, *Wired Wilderness.*

79   plastic resin: William W. Cochran and Rexford D. Lord, "A Radio-Tracking System for Wild Animals," *Journal of Wildlife Management* 27, no. 1 (1963), doi.org/10.2307/3797775.

79   skunks and racoons: Ibid.

79   stay in his spare room: Ibid.

80   impending nuclear war: Benson, *Wired Wilderness.*

81   "The space flights": Richard R. Graber, "Night Flight with a Thrush," *Audubon Magazine*, Nov.–Dec. 1965.

81   falcons: William W. Cochran and Roger D. Applegate, "Speed of Flapping Flight of Merlins and Peregrine Falcons," *Condor* 88, no. 3 (1986): 397–98, doi.org/10.2307/1368897.

81   eagles: Roger D. Applegate et al., "Observations of a Radio-Tagged Golden Eagle Terminating Fall Migration," *Journal of Raptor Research* 21, no. 2 (Summer 1987): 68–70.

81   According to one article: Jesse Greenspan, "Chasing Birds Across the Country . . . for Science," Audubon, Oct. 15, 2015, www.audubon.org/news/chasing-birds-across-countryfor-science.

81   As recently as 2015: Melissa S. Bowlin et al., "Unexplained Altitude Changes in a Migrating Thrush: Long-Flight Altitude Data from Radio-Telemetry," *Auk* 132, no. 4 (2015): 808–16, doi.org/10.1642/AUK -15-33.1.

81   "depletion of personal funds": William W. Cochran, "Following a Migrating Peregrine from Wisconsin to Mexico," *Hawk Chalk* 14, no. 2 (1975): 28–37.

82   Swainson's thrushes: Diane Evans Mack and Wang Yong, "Swainson's Thrush (*Catharus ustulatus*)," Birds of the World, Cornell Lab of Ornithology, birdsoftheworld.org/bow/species/swathr/cur/introduction.

84   total of 284 birds: Ana M. González, Nicholas J. Bayly, and Keith A. Hobson, "Earlier and Slower or Later and Faster: Spring Migration Pace Linked to Departure Time in a Neotropical Migrant Songbird," *Journal of Animal Ecology* 89, no. 12 (2020): 2840–51, doi.org/10.1111 /1365-2656.13359.

84   Their long journey home: Except where otherwise noted, details about the research project described in this section from author's interview and correspondence with González, June 2021.

85   inspired by Sputnik: Benson, *Wired Wilderness*.

85   fifty animals simultaneously: W. W. Cochran et al., "Automatic Radio-Tracking System for Monitoring Animal Movements," *BioScience* 15, no. 2 (1965): 98–100, doi.org/10.2307/1293346.

86   the Long Point array: Philip D. Taylor et al., "Landscape Movements of Migratory Birds and Bats Reveal an Expanded Scale of Stopover," *PLoS ONE* 6, no. 11 (2011): e27054, doi.org/10.1371/journal .pone.0027054.

86   Another array: Jill L. Deppe et al., "Fat, Weather, and Date Affect Migratory Songbirds' Departure Decisions, Routes, and Time It Takes to Cross the Gulf of Mexico," *Proceedings of the National Academy of Sciences* 112, no. 46 (2015): E6331–38, doi.org/10.1073/pnas.1503381112.

86   By the end of 2016: Philip Taylor et al., "The Motus Wildlife Tracking System: A Collaborative Research Network to Enhance the

Understanding of Wildlife Movement," *Avian Conservation and Ecology* 12, no. 1 (2017), doi.org/10.5751/ACE-00953-120108.

88   Chaplin Lake Motus tower: "Saskatchewan Migratory Shorebirds (#63)," Motus Wildlife Tracking System, accessed June 16, 2021, motus.org/data/project?id=63.

88   Incredibly, it was less: Ana M. González, "A Migratory Bird's Journey from the Andes of Colombia to North America: Leave Early and Take It Easy or Leave Late and Migrate Fast?," *Animal Ecology in Focus* (blog), Oct. 9, 2020, animalecologyinfocus.com/2020/10/09/a-migratory-birds-journey-from-the-andes-of-colombia-to-north-america-leave-early-and-take-it-easy-or-leave-late-and-migrate-fast/.

89   restoring the natural floodplain: "Why Reconnection Matters," Nature Conservancy, accessed Oct. 20, 2021, www.nature.org/en-us/about-us/where-we-work/united-states/illinois/stories-in-illinois/emiquon-reconnected-to-the-illinois-river/.

93   As of the summer of 2022: "By the Numbers," Motus Wildlife Tracking System, accessed June 8, 2022, motus.org/data/numbers.

93   including numerous birds: "Motus Species," Motus Wildlife Tracking System, accessed June 17, 2021, motus.org/data/species.

93   the network is growing all the time: "Motus Receiver Locations," Motus Wildlife Tracking System, accessed June 17, 2021, motus.org/data/receiversMap.

93   Caribbean Motus Collaboration: "The Caribbean Motus Collaboration—Help Us Develop This Exciting New Program!," *BirdsCaribbean* (blog), March 28, 2021, www.birdscaribbean.org/2021/03/the-caribbean-motus-collaboration-help-us-develop-this-exciting-new-program/.

Five | Higher, Further, Faster

95   put out a press release: "Bird Completes Epic Flight Across the Pacific," ScienceDaily, Sept. 17, 2007, www.sciencedaily.com/releases/2007/09/070915131205.htm.

95   curiously absent: Robert E. Gill Jr. et al., "Crossing the Ultimate
      Ecological Barrier: Evidence for an 11000-Km-Long Nonstop Flight
      from Alaska to New Zealand and Eastern Australia by Bar-Tailed
      Godwits," *Condor* 107, no. 1 (2005): 1–20, doi.org/10.1093/condor
      /107.1.1.

97   more than seven thousand miles: Robert E. Gill et al., "Extreme
      Endurance Flights by Landbirds Crossing the Pacific Ocean: Ecolog-
      ical Corridor Rather Than Barrier?," *Proceedings of the Royal Society B:
      Biological Sciences* 276, no. 1656 (2009): 447–57, doi.org/10.1098/rspb
      .2008.1142.

98   "we can put the receiver in orbit": Benson, *Wired Wilderness.*

98   a cow elk: Ben Goldfarb, "Monique the Space Elk and the Wild His-
      tory of Tracking Wildlife," *High Country News*, April 21, 2020, www.hcn
      .org/articles/wildlife-monique-the-space-elk-and-the-wild-history
      -of-tracking-wildlife.

98   ten million acres: Congressional Research Service, "Federal Land
      Ownership: Overview and Data," Feb. 21, 2020, fas.org/sgp/crs/misc
      /R42346.pdf.

99   Launched in 1978: Rebecca Morelle, "Argos: Keeping Track of the
      Planet," BBC News, June 7, 2007, news.bbc.co.uk/2/hi/science/nature
      /6701221.stm.

100   one can calculate: Ibid.

100   250 meters: This statistic according to an interview with Lance
      Jordan, Woods Hole Group.

100   movements of polar bears: Russel D. Andrews, "Tracking Marine
      Mammals," *Argos Forum*, Oct. 2009.

100   about six ounces: Thomas E. Strikwerda et al., "Bird-Borne Satellite
      Transmitter and Location Program," *Johns Hopkins APL Technical
      Digest* 7, no. 2 (1986): 203–8.

101   location fixes: Thomas E. Strikwerda et al., "The Bird-Borne Trans-
      mitter," *Johns Hopkins APL Technical Digest* 6, no. 1 (1985): 60–67.

101 "In order to capture a swan": Strikwerda et al., "Bird-Borne Satellite Transmitter and Location Program."

102 group of Russian ornithologists: P. S. Tomkovich et al., "First Indications of a Sharp Population Decline in the Globally Threatened Spoon-Billed Sandpiper *Eurynorhynchus pygmeus*," *Bird Conservation International* 12, no. 1 (March 2002): 1–18, doi.org/10.1017/S095927 0902002010.

102 340 individuals: Christoph Zöckler, Pyae Phyo Aung, and Sayam U. Chowdhury, "Summary of SBS Winter Counts 2021 and Proportion of Flagged Spoon-Billed Sandpiper," *Spoon-Billed Sandpiper Task Force News Bulletin*, May 2021.

105 identify a total: Q. Chang et al., "Post-Breeding Migration of Adult Spoon-Billed Sandpipers," *Wader Study* 127, no. 3 (2020): 200–209, doi .org/10.18194/ws.00201.

105 ten of the twenty-eight sites: Ibid.

108 thirty-one satellites: "Space Segment," GPS.gov, accessed July 2, 2021, www.gps.gov/systems/gps/space/.

108 atomic clock: "What Is an Atomic Clock?," NASA, accessed July 2, 2021, www.nasa.gov/feature/jpl/what-is-an-atomic-clock.

108 put that information together: "Satellite Navigation—GPS—How It Works," Federal Aviation Administration, accessed July 2, 2021, www .faa.gov/about/office_org/headquarters_offices/ato/service_units /techops/navservices/gnss/gps/howitworks/.

108 theory of relativity: Lee Mohon, "Einstein's Theory of Relativity, Vital for GPS, Seen in Distant Stars," NASA, Oct. 22, 2020, www.nasa .gov/mission_pages/chandra/images/einstein-s-theory-of-relativity -critical-for-gps-seen-in-distant-stars.html.

109 just after midnight: "Frequently Asked Questions About Selective Availability," GPS.gov, accessed July 2, 2021, www.gps.gov/systems /gps/modernization/sa/faq/.

109 "to make GPS more attractive": "Satellite Navigation—GPS—

Policy—Selective Availability," Federal Aviation Administration, Nov. 13, 2014, www.faa.gov/about/office_org/headquarters_offices /ato/service_units/techops/navservices/gnss/gps/policy/availability/.

109 ten times more accurate: "Brief History of GPS," Aerospace Corporation, accessed July 2, 2021, aerospace.org/article/brief-history-gps.

109 scientists in Germany: K. von Hünerbein et al., "A GPS-Based System for Recording the Flight Paths of Birds," *Naturwissenschaften* 87, no. 6 (2000): 278–79, doi.org/10.1007/s001140050721.

109 offshore foraging behavior: David Grémillet et al., "Offshore Diplomacy, or How Seabirds Mitigate Intra-specific Competition: A Case Study Based on GPS Tracking of Cape Gannets from Neighbouring Colonies," *Marine Ecology Progress Series* 268 (2004): 265–79, doi .org/10.3354/meps268265.

110 known about this split: Gary W. Page et al., "Annual Migratory Patterns of Long-Billed Curlews in the American West," *Condor* 116, no. 1 (2014): 50–61, doi.org/10.1650/CONDOR-12-185-R2.1.

114 only a day: Lucy A. Hawkes et al., "The Trans-Himalayan Flights of Bar-Headed Geese (*Anser indicus*)," *Proceedings of the National Academy of Sciences* 108, no. 23 (2011): 9516–19, doi.org/10.1073/pnas .1017295108.

114 geese didn't seem: L. A. Hawkes et al., "The Paradox of Extreme High-Altitude Migration in Bar-Headed Geese *Anser indicus*," *Proceedings of the Royal Society B: Biological Sciences* 280, no. 1750 (2013): 20122114, doi.org/10.1098/rspb.2012.2114.

114 manage sustained flight: Graham R. Scott et al., "How Bar-Headed Geese Fly over the Himalayas," *Physiology* 30, no. 2 (2015): 107–15, doi .org/10.1152/physiol.00050.2014.

114 Jessica Meir: Melanie Whiting, "Jessica U. Meir (Ph.D.) NASA Astronaut," NASA, Feb. 1, 2016, www.nasa.gov/astronauts/biographies /jessica-u-meir/biography.

116 magnetic compass: William W. Cochran, Henrik Mouritsen, and Martin Wikelski, "Migrating Songbirds Recalibrate Their Magnetic

Compass Daily from Twilight Cues," *Science* 304, no. 5669 (2004): 405–8, doi.org/10.1126/science.1095844.

116 bats navigate: Richard A. Holland et al., "Bat Orientation Using Earth's Magnetic Field," *Nature* 444, no. 7120 (Dec. 2006): 702, doi .org/10.1038/444702a.

116 remotely monitor: Silke S. Steiger et al., "Low Metabolism and In-active Lifestyle of a Tropical Rain Forest Bird Investigated via Heart-Rate Telemetry," *Physiological and Biochemical Zoology* 82, no. 5 (2009): 580–89, doi.org/10.1086/605336.

116 came online in 2003: "GPS and Camera Traps to Replace Radio Antennas in Tracking Animals on Barro Colorado Island," *Smithsonian Insider* (blog), Dec. 7, 2010, insider.si.edu/2010/12/gps-and-camera -traps-replace-radio-antennas-in-tracking-animals-on-barro-colorado -island/.

116 "It was not an isolated place": Sonia Shah, "How Far Does Wildlife Roam? Ask the 'Internet of Animals,'" *New York Times Magazine,* Jan. 12, 2021, www.nytimes.com/interactive/2021/01/12/magazine /animal-tracking-icarus.html.

117 Very Large Array: Daniel Dexter, "Remembering Professor Emeritus George Swenson Jr.," March 3, 2017, csl.illinois.edu/news/remembering -professor-emertius-george-swenson-jr.

117 the solution to Wikelski's problem: Shah, "How Far Does Wildlife Roam?"

117 still too large: Martin Wikelski et al., "Going Wild: What a Global Small-Animal Tracking System Could Do for Experimental Biolo-gists," *Journal of Experimental Biology* 210, no. 2 (2007): 181–86, doi .org/10.1242/jeb.02629.

118 controversial figure: Andrew Curry, "The Internet of Animals That Could Help to Save Vanishing Wildlife," *Nature* 562, no. 7727 (2018): 322–26, doi.org/10.1038/d41586-018-07036-2.

118 pushed back: Ibid.

118 supposed to be available: "Icarus Is Launched," Icarus, March 10, 2020, www.icarus.mpg.de/90671/news_publication_14575703_trans ferred.

118 abruptly stopped transmitting: Elizabeth Pennisi, "War Halts Project to Track Wildlife from Space," Science, March 24, 2022, accessed April 18, 2022, www.science.org/content/article/war-halts-project -track-wildlife-space.

118 fifth of an ounce: "Transmitters," Icarus, accessed July 20, 2021, www.icarus.mpg.de/28874/sensor-animals-tracking.

118 €500 each: Jim Robbins, "With an Internet of Animals, Scientists Aim to Track and Save Wildlife," New York Times, June 9, 2020, www .nytimes.com/2020/06/09/science/space-station-wildlife.html.

118 five times as much: "MTI: Avian Transmitter Pricing," Microwave Telemetry Inc., accessed July 20, 2021, www.microwavetelemetry.com /avian_transmitter_pricing.

119 tiny "nanosats": "ANGELS, Successful Launch: The Argos Metamorphosis Is on Its Way," Argos, Dec. 18, 2019, www.argos-system.org /angels-the-first-argos-nanosat-is-here/.

Six | Navigating by the Sun

120 Habitat loss: David W. Winkler, Shawn M. Billerman, and Irby J. Lovette, "New World Warblers (Parulidae), Version 1.1," Birds of the World, Cornell Lab of Ornithology, doi.org/10.2173/bow.paruli 1.01.1.

121 hummingbirds: Theodore J. Zenzal et al., "Migratory Hummingbirds Make Their Own Rules: The Decision to Resume Migration Along a Barrier," Animal Behaviour 137 (March 2018): 215–24, doi.org /10.1016/j.anbehav.2018.01.019.

122 "dead reckoning": Rory P. Wilson et al., "Determination of Movements of African Penguins (Spheniscus demersus) Using a Compass System: Dead Reckoning May Be an Alternative to Telemetry," Journal

of Experimental Biology 157, no. 1 (1991): 557–64, doi.org/10.1242
/jeb.157.1.557.

122 "Wilson recalled": David Grémillet, "Let There Be Light—My Personal Account of How Rory P. Wilson Invented Seabird Geolocation,"
Oct. 2015, dx.doi.org/10.13140/RG.2.1.1731.9123.

123 "top conservation heroes": Kevin Sullivan, "Animal Movement
Expert Rory Wilson Named as One of BBC's Top 50 Conservation
Heroes," Swansea University, May 11, 2015, www-2018.swansea.ac.uk
/press-office/news-archive/2015/animalmovementexpertrorywilso
nnamedasoneofbbcstop50conservationheroes.php.

123 the 1530s: A. Pogo, "Gemma Frisius, His Method of Determining
Differences of Longitude by Transporting Timepieces (1530), and His
Treatise on Triangulation (1533)," *Isis* 22, no. 2 (1935): 469–506, doi.org
/10.1086/346920.

123 first published study: Rory P. Wilson et al., "Foraging Areas of
Magellanic Penguins *Spheniscus magellanicus* Breeding at San Lorenzo,
Argentina, During the Incubation Period," *Marine Ecology Progress Series*
129, no. 1/3 (1995): 1–6, www.jstor.org/stable/24855568.

124 "extravagant": Grémillet, "Let There Be Light."

124 "genius": Christine McGourty, "It Will Be All White—Eventually,"
BBC News, Dec. 22, 2001, news.bbc.co.uk/2/hi/programmes/from
_our_own_correspondent/1724627.stm.

124 "idiosyncratic," and "basically an inventor manqué": From an email
to the author from John Croxall, Aug. 20, 2021.

124 "would prowl silently": McGourty, "It Will Be All White—
Eventually."

125 passed away: "The Obituary Notice for Vsevolod Afanasyev," Funeral Notices, Dec. 1, 2018, funeral-notices.co.uk/notice/AFANASYEV
/4553585.

126 Ian Nisbet: Nisbet, "Autumn Migration of the Blackpoll Warbler."

126 Bertram Murray: Bertram G. Murray Jr., "A Critical Review of the Transoceanic Migration of the Blackpoll Warbler," *Auk* 106, no. 1 (1989): 8–17, doi.org/10.2307/4087751.

127 spent his PhD years: William V. DeLuca, "Ecology and Conservation of the Montane Forest Avian Community in Northeastern North America" (PhD diss., University of Massachusetts Amherst, 2013), www.proquest.com/openview/fe4723afe9098f2f09b6dd28d4d4d135/1?pq-origsite=gscholar&cbl=18750.

127 study of the ovenbird: Michael T. Hallworth et al., "Migratory Connectivity of a Neotropical Migratory Songbird Revealed by Archival Light-Level Geolocators," *Ecological Applications* 25, no. 2 (2015): 336–47, doi.org/10.1890/14-0195.1.

129 The longest single transoceanic flight: William V. DeLuca et al., "Transoceanic Migration by a 12 g Songbird," *Biology Letters* 11, no. 4 (2015): 20141045, doi.org/10.1098/rsbl.2014.1045.

129 a follow-up study: William V. DeLuca et al., "A Boreal Songbird's 20,000 Km Migration Across North America and the Atlantic Ocean," *Ecology* 100, no. 5 (2019): e02651, doi.org/10.1002/ecy.2651.

129 a mistake: Douglas B. McNair and Ian C. T. Nisbet, "Status and Abundance of Blackpoll Warblers in Autumn on the Coast of the Southeastern United States: An Update," *Southeastern Naturalist* 19, no. 2 (2020): 241–55, doi.org/10.1656/058.019.0205.

130 George Lowery's backyard: Interview with Sidney Gauthreaux.

134 fledgling golden-winged warblers: H. M. Streby et al., "Post-independence Fledgling Ecology in a Migratory Songbird: Implications for Breeding-Grounds Conservation," *Animal Conservation* 18, no. 3 (2015): 228–35, doi.org/10.1111/acv.12163.

134 Yet another researcher: Jared D. Wolfe and Erik I. Johnson, "Geolocator Reveals Migratory and Winter Movements of a Prothonotary Warbler," *Journal of Field Ornithology* 86, no. 3 (2015): 238–43, doi.org/10.1111/jofo.12107.

134 more than forty geolocators: Gunnar R. Kramer et al., "Population Trends in *Vermivora* Warblers Are Linked to Strong Migratory Connectivity," *Proceedings of the National Academy of Sciences* 115, no. 14 (2018): E3192–200, doi.org/10.1073/pnas.1718985115.

136 twenty-nine Connecticut warblers: E. A. McKinnon, C. Artuso, and O. P. Love, "The Mystery of the Missing Warbler," *Ecology* 98, no. 7 (2017): 1970–72, doi.org/10.1002/ecy.1844.

136 seven miles: Adam M. Fudickar, Martin Wikelski, and Jesko Partecke, "Tracking Migratory Songbirds: Accuracy of Light-Level Loggers (Geolocators) in Forest Habitats," *Methods in Ecology and Evolution* 3, no. 1 (2012): 47–52, doi.org/10.1111/j.2041-210X.2011.00136.x.

136 three hundred miles: Eldar Rakhimberdiev et al., "Comparing Inferences of Solar Geolocation Data Against High-Precision GPS Data: Annual Movements of a Double-Tagged Black-Tailed Godwit," *Journal of Avian Biology* 47, no. 4 (2016): 589–96, doi.org/10.1111/jav.00891.

137 first migratory songbirds: Bridget J. M. Stutchbury et al., "Tracking Long-Distance Songbird Migration by Using Geolocators," *Science* 323, no. 5916 (2009): 896, doi.org/10.1126/science.1166664.

137 phalaropes: Malcolm Smith et al., "Geolocator Tagging Reveals Pacific Migration of Red-Necked Phalarope *Phalaropus lobatus* Breeding in Scotland," *Ibis* 156, no. 4 (2014): 870–73, doi.org/10.1111/ibi.12196.

137 common rosefinches: Simeon Lisovski et al., "The Indo-European Flyway: Opportunities and Constraints Reflected by Common Rosefinches Breeding Across Europe," *Journal of Biogeography* 48, no. 6 (2021): 1255–66, doi.org/10.1111/jbi.14085.

138 painted buntings: Clark S. Rushing et al., "Integrating Tracking and Resight Data Enables Unbiased Inferences About Migratory Connectivity and Winter Range Survival from Archival Tags," *Ornithological Applications* 123, no. 2 (2021), doi.org/10.1093/ornithapp/duab010.

138 small but measurable: Vojtěch Brlík et al., "Weak Effects of Geolocators on Small Birds: A Meta-analysis Controlled for Phylogeny and Publication Bias," *Journal of Animal Ecology* 89, no. 1 (2020): 207–20, doi.org/10.1111/1365-2656.12962.

138 common yellowthroats: Conor C. Taff et al., "Geolocator Deployment Reduces Return Rate, Alters Selection, and Impacts Demography in a Small Songbird," *PLoS ONE* 13, no. 12 (2018): e0207783, doi .org/10.1371/journal.pone.0207783.

139 just a suggestion: Douglas G. Barron, Jeffrey D. Brawn, and Patrick J. Weatherhead, "Meta-analysis of Transmitter Effects on Avian Behaviour and Ecology," *Methods in Ecology and Evolution* 1, no. 2 (2010): 180–87, doi.org/10.1111/j.2041-210X.2010.00013.x.

139 "there was no device mass threshold": Graham R. Geen, Robert A. Robinson, and Stephen R. Baillie, "Effects of Tracking Devices on Individual Birds—a Review of the Evidence," *Journal of Avian Biology* 50, no. 2 (2019), doi.org/10.1111/jav.01823.

140 smaller and smaller birds: Steven J. Portugal and Craig R. White, "Miniaturization of Biologgers Is Not Alleviating the 5% Rule," *Methods in Ecology and Evolution* 9, no. 7 (2018): 1662–66, doi.org/10.1111 /2041-210X.13013.

Seven | You Are Where You Eat

143 one in about sixty-four hundred: R. Hagemann, G. Nief, and E. Roth, "Absolute Isotopic Scale for Deuterium Analysis of Natural Waters. Absolute D/H Ratio for SMOW," *Tellus* 22, no. 6 (1970): 712–15, doi.org/10.3402/tellusa.v22i6.10278.

145 Radiocarbon dating: Rachel Wood, "Explainer: What Is Radiocarbon Dating and How Does It Work?," The Conversation, Nov. 27, 2012, theconversation.com/explainer-what-is-radiocarbon-dating-and-how -does-it-work-9690.

146 ancient human populations: Jonathon E. Ericson, "Strontium Isotope Characterization in the Study of Prehistoric Human Ecology," *Journal of Human Evolution* 14, no. 5 (1985): 503–14, doi.org/10.1016/S0047-2484 (85)80029-4.

146 poached elephant ivory: N. J. van der Merwe et al., "Source-Area Determination of Elephant Ivory by Isotopic Analysis," *Nature* 346, no. 6286 (1990): 744–46, doi.org/10.1038/346744a0.

146 narwhals and polar bears: Keith A. Hobson and Harold E. Welch, "Determination of Trophic Relationships Within a High Arctic Marine Food Web Using δ 13 C and δ 15 N Analysis," *Marine Ecology Progress Series* 84, no. 1 (1992): 9–18, www.jstor.org/stable/24829721.

146 a "latitudinal gradient": W. Dansgaard, "Stable Isotopes in Precipitation," *Tellus* 16, no. 4 (1964): 436–68, doi.org/10.3402/tellusa.v16i4.8993.

148 For every feather: K. A. Hobson and Leonard I. Wassenaar, "Linking Breeding and Wintering Grounds of Neotropical Migrant Songbirds Using Stable Hydrogen Isotopic Analysis of Feathers," *Oecologia* 109, no. 1 (1996): 142–48, doi.org/10.1007/s004420050068.

148 Page Chamberlain: C. P. Chamberlain et al., "The Use of Isotope Tracers for Identifying Populations of Migratory Birds," *Oecologia* 109, no. 1 (1996): 132–41, doi.org/10.1007/s004420050067.

148 American crow: Andrea K. Townsend et al., "Where Do Winter Crows Go? Characterizing Partial Migration of American Crows with Satellite Telemetry, Stable Isotopes, and Molecular Markers," *Auk* 135, no. 4 (2018): 964–74, doi.org/10.1642/AUK-18-23.1.

148 Vaux's swift: Ellis L. Smith et al., "Breeding Origins and Migratory Connectivity at a Northern Roost of Vaux's Swift, a Declining Aerial Insectivore," *Condor* 121, no. 3 (2019), doi.org/10.1093/condor/duz034.

152 Kestrel populations: "American Kestrel Population Decline," American Kestrel Partnership, accessed Nov. 9, 2021, kestrel.peregrinefund.org/decline.

153 tissues of monarchs: Leonard I. Wassenaar and Keith A. Hobson, "Natal Origins of Migratory Monarch Butterflies at Wintering Colonies in Mexico: New Isotopic Evidence," *Proceedings of the National Academy of Sciences* 95, no. 26 (1998): 15436–39, doi.org/10.1073/pnas.95.26.15436.

153 Monsanto: "Company History," Monsanto, April 23, 2008, web.archive.org/web/20080423174556/http://www.monsanto.com/who_we_are/history.asp.

153 more than 80 percent: "Saving the Monarch Butterfly," Center for Biological Diversity, accessed Oct. 20, 2021, www.biologicaldiversity .org/species/invertebrates/monarch_butterfly/.

157 "Solving a Migration Riddle": Keith A. Hobson et al., "Solving a Migration Riddle Using Isoscapes: House Martins from a Dutch Village Winter over West Africa," *PLoS ONE* 7, no. 9 (2012): e45005, doi .org/10.1371/journal.pone.0045005.

159 results were dramatic: Laura Cárdenas-Ortiz et al., "Defining Catchment Origins of a Geographical Bottleneck: Implications of Population Mixing and Phenological Overlap for the Conservation of Neotropical Migratory Birds," *Condor* 122, no. 2 (2020), doi .org/10.1093/condor/duaa004.

161 Colombia Resurvey Project: "About," Colombia Resurvey Project, accessed Oct. 14, 2021, colombiaresurveyproject.com/about/.

161 rusty blackbirds: Keith A. Hobson et al., "Migratory Connectivity in the Rusty Blackbird: Isotopic Evidence from Feathers of Historical and Contemporary Specimens," *Condor* 112, no. 4 (2010): 778–88, doi .org/10.1525/cond.2010.100146.

162 named in his honor: Felipe Guhl, "Cornelis Johannes Marinkelle, biólogo, médico, Ph.D.," *Infectio* 16, no. 1 (2012), www.revistainfectio .org/index.php/infectio/article/view/506.

162 a staggering rate: John R. Sauer et al., "The First 50 Years of the North American Breeding Bird Survey," *Condor* 119, no. 3 (2017): 576–93, doi.org/10.1650/CONDOR-17-83.1.

162 scientists forecast: William DeLuca et al., "Blackpoll Warbler (*Setophaga striata*)," Birds of the World, Cornell Lab of Ornithology, birdsoftheworld.org/bow/species/bkpwar/cur/introduction.

163 more than 370 miles: Camila Gómez et al., "Migratory Connectivity Then and Now: A Northward Shift in Breeding Origins of a Long-Distance Migratory Bird Wintering in the Tropics," *Proceedings of the Royal Society B: Biological Sciences* 288, no. 1948 (2021), doi.org/10.1098 /rspb.2021.0188.

Eight | The Feather Library

166 ice covered: "How Does Present Glacier Extent and Sea Level Compare to the Extent of Glaciers and Global Sea Level During the Last Glacial Maximum (LGM)?," USGS, accessed Dec. 1, 2021, www .usgs.gov/faqs/how-does-present-glacier-extent-and-sea-level-compare -extent-glaciers-and-global-sea-level.

166 then expanded again: Scott K. Johnson, "Did Birds Still Migrate During Ice Ages?," *Ars Technica*, Feb. 18, 2020, arstechnica.com/science /2020/02/did-birds-still-migrate-during-ice-ages/.

167 invented a tool: R. Jefferson Smith and Thomas Bates Smith, "Portable Device for Measuring Seed Hardness," *Journal of Field Ornithology* 60, no. 1 (Winter 1989): 56–59.

167 two distinct varieties: Thomas Bates Smith, "Bill Size Polymorphism and Intraspecific Niche Utilization in an African Finch," *Nature* 329, no. 6141 (1987): 717–19, doi.org/10.1038/329717a0.

169 ancient programming kicks in: Kristen C. Ruegg and Thomas B. Smith, "Not as the Crow Flies: A Historical Explanation for Circuitous Migration in Swainson's Thrush (*Catharus ustulatus*)," *Proceedings of the Royal Society B: Biological Sciences* 269, no. 1498 (2002): 1375–81, doi.org /10.1098/rspb.2002.2032.

170 billion or more: Guojie Zhang et al., "Comparative Genomics Reveals Insights into Avian Genome Evolution and Adaptation," *Science* 346, no. 6215 (2014): 1311–20, doi.org/10.1126/science.1251385.

170 around twenty thousand: Nan Xu et al., "The Mitochondrial Genome and Phylogenetic Characteristics of the Thick-Billed Green-Pigeon, *Treron curvirostra*: The First Sequence for the Genus," *ZooKeys* 1041 (2021): 167–82, doi.org/10.3897/zookeys.1041.60150.

170 Mari Kimura: M. Kimura et al., "Phylogeographical Approaches to Assessing Demographic Connectivity Between Breeding and Overwintering Regions in a Nearctic–Neotropical Warbler (*Wilsonia pusilla*)," *Molecular Ecology* 11, no. 9 (2002): 1605–16, doi.org/10.1046 /j.1365-294X.2002.01551.x.

170 variations in small sections: Darren E. Irwin, Jessica H. Irwin, and Thomas B. Smith, "Genetic Variation and Seasonal Migratory Connectivity in Wilson's Warblers (*Wilsonia pusilla*): Species-Level Differences in Nuclear DNA Between Western and Eastern Populations," *Molecular Ecology* 20, no. 15 (2011): 3102–15, doi.org/10.1111/j.1365-294 X.2011.05159.x.

171 experimenting with combining: Sonya M. Clegg et al., "Combining Genetic Markers and Stable Isotopes to Reveal Population Connectivity and Migration Patterns in a Neotropical Migrant, Wilson's Warbler (*Wilsonia pusilla*)," *Molecular Ecology* 12, no. 4 (2003): 819–30, doi.org/10.1046/j.1365-294X.2003.01757.x.

171 three billion dollars: "Human Genome Project FAQ," Genome.gov, accessed Dec. 2, 2021, www.genome.gov/human-genome-project /Completion-FAQ.

171 huge leaps: Sarita Sonavane, "Lessons from the Human Genome Project," *Science in the News* (blog), Feb. 27, 2019, sitn.hms.harvard.edu /flash/2019/lessons-from-the-human-genome-project/.

172 Crucially, they invented: Leroy Hood and Lee Rowen, "The Human Genome Project: Big Science Transforms Biology and Medicine," *Genome Medicine* 5, no. 9 (2013): 79, doi.org/10.1186/gm483.

172 a tedious process: "How Does DNA Sequencing Work?," Genome News Network, accessed Jan. 4, 2022, www.genomenewsnetwork.org /resources/whats_a_genome/Chp2_2.shtml.

177 more than 1,600 birds: Kristen C. Ruegg et al., "Mapping Migration in a Songbird Using High-Resolution Genetic Markers," *Molecular Ecology* 23, no. 23 (2014): 5726–39, doi.org/10.1111/mec.12977.

179 an ornithology conference: Susan K. Skagen and Sara Oyler-McCance, *AOU-COS-SCO 2014 Joint Meeting Program* (Estes Park, Colo.: American Ornithologists' Union, Cooper Ornithological Society, and Society of Canadian Ornithologists, 2014), americanornithology.org /wp-content/uploads/2019/08/2014_AOU_COS_SCO_Program.pdf.

179 American kestrel: Kristen C. Ruegg et al., "The American Kestrel

(*Falco sparverius*) Genoscape: Implications for Monitoring, Management, and Subspecies Boundaries," *Ornithology* 138, no. 2 (2021): ukaa051, doi.org/10.1093/auk/ukaa051.

181 229 individual yellow warblers: Rachael A. Bay et al., "Genomic Signals of Selection Predict Climate-Driven Population Declines in a Migratory Bird," *Science* 359, no. 6371 (2018): 83–86, doi.org/10.1126/science.aan4380.

181 already declining faster: Ibid.

182 follow-up study: Rachael A. Bay et al., "Genetic Variation Reveals Individual-Level Climate Tracking Across the Annual Cycle of a Migratory Bird," *Ecology Letters* 24, no. 4 (2021): 819–28, doi.org/10.1111/ele.13706.

182 limited conservation funding: Kristen C. Ruegg et al., "A Genoscape-Network Model for Conservation Prioritization in a Migratory Bird," *Conservation Biology* 34, no. 6 (2020): 1482–91, doi.org/10.1111/cobi.13536.

Nine | Vox Populi

185 independently coined: Hauke Riesch and Clive Potter, "Citizen Science as Seen by Scientists: Methodological, Epistemological, and Ethical Dimensions," *Public Understanding of Science* 23, no. 1 (2014): 107–20, doi.org/10.1177/0963662513497324.

185 political connotations: "Why We're Changing from 'Citizen Science' to 'Community Science,'" Audubon Center at Debs Park, May 3, 2018, debspark.audubon.org/news/why-were-changing-citizen-science-community-science.

185 "Until comparatively recently": Mary H. Clench, "The Importance of Contributions by Amateurs to American Ornithology: A Story History," *Bird Observer* 19, no. 2 (1991): 67–74, sora.unm.edu/node/141914.

186 first Christmas Bird Count: "History of the Christmas Bird Count," Audubon, Jan. 21, 2015, www.audubon.org/conservation/history-christmas-bird-count.

186 In December 2019: Geoff LeBaron, "120th Christmas Bird Count Summary," Audubon, Dec. 7, 2020, www.audubon.org/news/120th -christmas-bird-count-summary.

186 Modern Christmas Bird Counts: "Answers to Your Top Questions About the Christmas Bird Count," Audubon, Sept. 24, 2021, www .audubon.org/answers-your-top-questions-about-christmas-bird -count.

187 Journey North: "Hummingbirds," Journey North, accessed Jan. 21, 2022, journeynorth.org/hummingbirds.

187 Project FeederWatch: "About," Project FeederWatch, accessed Jan. 21, 2022, feederwatch.org/about/.

187 Great Backyard Bird Count: "About," Great Backyard Bird Count, accessed Jan. 21, 2022, www.birdcount.org/about/.

187 Nightjar Survey Network: "About the Nightjar Survey Network," Nightjar Survey Network, 2022, www.nightjars.org/about/about-the -nightjar-survey-network/.

187 International Shorebird Survey: Brad Winn, "International Shore- bird Survey," Manomet, accessed Feb. 3, 2022, www.manomet.org /project/international-shorebird-survey/.

187 iNaturalist: "About," iNaturalist, accessed Jan. 24, 2022, www .inaturalist.org/pages/about.

187 folding proteins: "The Science Behind Foldit," Foldit, accessed Jan. 24, 2022, fold.it/portal/info/about.

187 intelligent life: "About SETI@home," SETI@home, accessed Jan. 24, 2022, setiathome.berkeley.edu/sah_about.php.

188 "Birds both inspire and permit": Philip C. Stouffer, "Celebrating the North American Breeding Bird Survey," Oxford Academic, 2017, academic.oup.com/aosjournals/pages/breeding_bird_survey.

188 as a toddler: "Chandler Robbins," USGS, March 23, 2017, www .usgs.gov/news/national-news-release/chandler-robbins-inspired -generations-scientists-and-birders-1918-2017.

188 In 1940, he graduated: Jay M. Sheppard, Deanna K. Dawson, and John R. Sauer, "Chandler S. Robbins, 1918–2017," *Auk* 134, no. 4 (2017): 935–38, doi.org/10.1642/AUK-17-140.1.

189 Multiple threads: Sauer et al., "First 50 Years of the North American Breeding Bird Survey."

189 "I had to write her back": "Chandler Robbins Oral History Transcript," USFWS National Digital Library, July 1, 2008, digitalmedia .fws.gov/digital/collection/document/id/1214.

190 In 1966, after a year: Sheppard, Dawson, and Sauer, "Chandler S. Robbins, 1918–2017."

190 had to hustle: Per interview with David Ziolkowski, Jan. 21, 2022.

190 skeptical of the methodology: Sauer et al., "First 50 Years of the North American Breeding Bird Survey."

190 His own wife: "Chandler Robbins Oral History Transcript."

190 two years later: J. R. Sauer et al., "The North American Breeding Bird Survey," USGS, 1997, www.mbr-pwrc.usgs.gov/bbs/genintro.html.

190 more than four thousand: "About BBS," USGS Patuxent Wildlife Research Center, accessed Jan. 26, 2022, www.pwrc.usgs.gov/bbs/about/.

190 a little more than half: Ziolkowski interview.

191 The only requirements: "Participating in the North American Breeding Bird Survey," USGS Patuxent Wildlife Research Center, 2018, www.pwrc.usgs.gov/bbs/participate/.

191 half an hour before sunrise: "About BBS."

191 permanent physical record: Ziolkowski interview.

193 sounded the first alarm: C. S. Robbins et al., "Population Declines in North American Birds That Migrate to the Neotropics," *Proceedings of the National Academy of Sciences* 86, no. 19 (1989): 7658–62, doi.org /10.1073/pnas.86.19.7658.

194 Laysan albatross named Wisdom: Darryl Fears, "Albatross Named Wisdom Astounds Scientists by Producing Chick at Age 62," *Washington Post*, Feb. 5, 2013, www.washingtonpost.com/national/health -science/albatross-named-wisdom-astounds-scientists-by-giving -birth-at-age-62/2013/02/05/f46a68a6-6fc5-11e2-8b8d-e0b59a1b8 e2a_story.html.

195 "rays of sunshine": "Chandler Robbins Oral History Transcript."

195 continued to come: Sheppard, Dawson, and Sauer, "Chandler S. Robbins, 1918–2017."

195 "the Bruce Springsteen of birding": Darryl Fears, "This Humble Scientist, a 'National Treasure,' Showed Us How to Understand Birds," *Washington Post*, March 24, 2017, www.washingtonpost.com/news /animalia/wp/2017/03/24/this-humble-scientist-a-national-treasure -showed-us-how-to-understand-birds/.

196 launched BirdSource: "'Birdsource' Website for Citizen-Science Data," Newswise, Feb. 14, 1997, www.newswise.com/articles/bird source-website-for-citizen-science-data.

197 six core pieces: John W. Fitzpatrick et al., "Introducing eBird: The Union of Passion and Purpose," *North American Birds* 56, no. 1 (2002): 11–12, sora.unm.edu/node/113636.

199 took ten years: Hugh Powell, "EBird and a Hundred Million Points of Light," *Living Bird*, Jan. 15, 2015, www.allaboutbirds.org/news/a -hundred-million-points-of-light/.

199 more than 700,000 birders: "2021 Year in Review: EBird, Merlin, Macaulay Library, and Birds of the World," eBird, Dec. 21, 2021, ebird .org/ebird/news/2021-year-in-review.

199 herald petrel: Chris Daly (@Liamdaly620), "I . . . question this iden-tification," Twitter, Feb. 9, 2022, 4:40 p.m., twitter.com/Liamdaly620 /status/1491572778285375492.

200 different routes: Frank A. La Sorte et al., "The Role of Atmospheric Conditions in the Seasonal Dynamics of North American Migration

Flyways," *Journal of Biogeography* 41, no. 9 (2014): 1685–96, doi.org /10.1111/jbi.12328.

200 Radar ornithologists: Kyle G. Horton et al., "Holding Steady: Little Change in Intensity or Timing of Bird Migration over the Gulf of Mexico," *Global Change Biology* 25, no. 3 (2019): 1106–18, doi.org/10 .1111/gcb.14540.

200 common potoo: Lucas W. DeGroote et al., "Citizen Science Data Reveals the Cryptic Migration of the Common Potoo *Nyctibius griseus* in Brazil," *Ibis* 163, no. 2 (2021): 380–89, doi.org/10.1111/ibi.12904.

200 Less traffic: Michael B. Schrimpf et al., "Reduced Human Activity During COVID-19 Alters Avian Land Use Across North America," *Science Advances*, Sept. 2021, doi.org/10.1126/sciadv.abf5073.

200 closures of many parks: Wesley M. Hochachka et al., "Regional Variation in the Impacts of the COVID-19 Pandemic on the Quantity and Quality of Data Collected by the Project eBird," *Biological Conservation* 254 (Feb. 2021), doi.org/10.1016/j.biocon.2021.108974.

202 migratory bottleneck: William V. DeLuca et al., "The Colorado River Delta and California's Central Valley Are Critical Regions for Many Migrating North American Landbirds," *Ornithological Applications* 123, no. 1 (2021), doi.org/10.1093/ornithapp/duaa064.

Conclusion | Sky Full of Hope

207 In 2019, a massive new study: Kenneth V. Rosenberg et al., "Decline of the North American Avifauna," *Science* 366, no. 6461 (2019): 120–24, doi.org/10.1126/science.aaw1313.

208 actions they could take: "7 Simple Actions," #BringBirdsBack, accessed Feb. 28, 2022, www.3billionbirds.org/7-simple-actions.

208 exposure to the pesticide: Brian Woodbridge, Karen K. Finley, and S. Trent Seager, "An Investigation of the Swainson's Hawk in Argentina," *Journal of Raptor Research* 29, no. 3 (1995): 202–4, sora.unm.edu /node/53483.

208 made a plan: Les Line, "Accord Is Reached to Recall Pesticide Devastating Hawk," *New York Times*, Oct. 15, 1996, www.nytimes
.com/1996/10/15/science/accord-is-reached-to-recall-pesticide
-devastating-hawk.html.

208 More recently, land was added: Mikko Jimenez, "A Tale of Two Migration Routes: How Prothonotary Warblers Make Their Way Home,"
Audubon, March 25, 2021, www.audubon.org/news/a-tale-two
-migration-routes-how-prothonotary-warblers-make-their-way-home.

208 newly identified hot spot: Dean Russell, "In the Middle of the Atlantic, an Overlooked Seabird Hotspot," *Hakai Magazine*, Nov. 22, 2021,
hakaimagazine.com/news/in-the-middle-of-the-atlantic-an-over
looked-seabird-hotspot/.

210 colleagues have identified: "Road to Recovery Urgency Lists,"
Marra Lab, 2022, marralab.com/r2r-urgency-list/.

Photo Credits

- *Insert page 4, bottom left:* © 2021, Eli Bridge

- *Insert page 4, bottom right:* © 2021, Adéla Minařík

- *Insert page 5, top left:* © 2021, Evan Heisman

- *Insert page 5, top right:* © 2021, Rebecca Heisman

- *Insert page 5, bottom:* © 2021, Rebecca Heisman

- *Insert page 6, top:* © 2019, Eric Rasmussen

- *Insert page 6, bottom:* © 2022, Kyle Horton

- *Insert page 7, top:* © 2020, Kyle Horton

- *Insert page 7, bottom:* © 1996, Carroll Belser

- *Insert page 8, top:* Image provided courtesy of Illinois Natural History Survey of the University of Illinois at Urbana-Champaign Prairie Research Institute

- *Insert page 8, middle:* © 2019, Ana González

- *Insert page 8, bottom:* © 2015, Ana González

- *Insert page 9, top left:* © 2021, Rebecca Heisman

- *Insert page 9, top right:* © 2021, Rebecca Heisman

- *Insert page 9, middle:* © 2021, Rebecca Heisman

- *Insert page 9, bottom:* © 2021, Scott Whittle

- *Insert page 10, top left:* 2007, Daniel Ruthrauff, courtesy of USGS

- *Insert page 10, top right:* © 2016, Ewan Weston

- *Insert page 10, middle left:* © 2021, Rebecca Heisman

- *Insert page 10, middle right:* © 2021, Rebecca Heisman

- *Insert page 10, bottom:* © 2011, Lucy Hawkes

- *Insert page 11, top:* © 2016, Emily McKinnon

- *Insert page 11, middle:* © 2016, Emily McKinnon

- *Insert page 11, bottom left:* © 1994, Rory Wilson

- *Insert page 11, bottom right:* © 2016, Gunnar Kramer

- *Insert page 12, top left:* © 2015, Gunnar Kramer

- *Insert page 12, top right:* © 2022, Rebecca Heisman

- *Insert page 12, bottom:* © 2022, Rebecca Heisman

- *Insert page 13, top:* © 2021, Rebecca Heisman

- *Insert page 13, second from top:* © 2021, Rebecca Heisman

- *Insert page 13, third from top:* © 2021, Rebecca Heisman

- *Insert page 13, bottom:* © 2021, Rebecca Heisman

- *Insert page 14, top:* © 2021, Theunis Piersma

- *Insert page 14, middle:* © 2015, Benjamin Dudek

- *Insert page 14, bottom:* © 2021, Carolina Rojas Céspedes

- *Insert page 15, top:* © 2022, Christine Rayne

- *Insert page 15, bottom:* © 2022, Christine Rayne

- *Insert page 16, top:* Barbara Dowell, courtesy of USGS

- *Insert page 16, middle:* © 2022, Rebecca Heisman

Index

About the Author

REBECCA HEISMAN is a science writer based in eastern Washington who loves nerding out about birds. She's contributed to publications including *Audubon, Sierra, Hakai Magazine, bioGraphic, Living Bird* (the magazine of the Cornell Lab of Ornithology), and *Bird Conservation* (the magazine of the American Bird Conservancy), and has worked for the American Ornithological Society (AOS), the world's largest professional organization for bird scientists.